# New Explorations in the Economics of Technical Change

# New Explorations in the Economics of Technical Change

Edited by
C. Freeman and L. Soete

Pinter Publishers, London & New York

First published in Great Britain in 1990 by
Pinter Publishers Limited
25 Floral Street, London WC2E 9DS and PO Box 197
Irvington, NY 10533



**British Library Cataloguing in Publication Data**

A CIP catalogue record for this book is available from the
British Library
ISBN 0 86187 128 6

**Library of Congress Cataloging-in-Publication Data**
A CIP record for this book is available from
the Library of Congress

Typeset by DP Photosetting, Aylesbury, Bucks
Printed and bound in Great Britain by Biddles Ltd.

# Contents

# Introduction

*Christopher Freeman and Luc Soete*

This book is based on a set of papers prepared for a Conference held at the University of Limburg in November 1989. All of the chapters in the book, except for one invited chapter by a Japanese colleague, are based on research undertaken over the last year or two at the Maastricht Economic Research institute on Innovation and Technology (MERIT). Although the scope of the chapters is fairly broad, it will become apparent that they are in fact closely interrelated despite the diversity of approach.

The chapters are grouped in three parts. The first group deals with problems of *innovation* management and strategy of the firm. The second group deals with the *diffusion* of innovations within a major service industry (banking) and a major manufacturing industry (automobiles), and with the organisational changes which accompany technical innovation, whether in banking or in the car industry. The third group is concerned with basic *theory* and theoretical models and with policy.

In the first part, the chapter by Hagedoorn and Schakenraad takes up an interesting and controversial issue in the study of innovation: why do firms apparently co-operate much more today in developing the new technologies; such as biotechnology, information technology and materials technology? Using a databank of over 7,000 co-operative arrangements in these advanced technologies, the authors show that one of the commonest explanations of 'networking' — sharing of risks and costs — is not in fact a dominant motive. Far more important are strategic considerations and complementarities in technology.

The second chapter, by Romme, also suggests serious limits on the value of transaction cost theory in relation to innovation and the importance of strategic behaviour of the firm. It develops an original 'self-organisation' theory of strategy formation and further develops the notion of 'repertoires' of behavioural routines enriching the original Nelson and Winter conception. The example of the failure of FACIT to move from electro-mechanical to electronic technology provides a nice illustration of Romme's theory.

The next two chapters both deal with the problems of managing innovation in the chemical industry. Chapter 3, by Cobbenhagen, den Hertog and Philips, is based on a project studying the real life problems of managing process innovations in several firms in the Netherlands. It concludes that contemporary theory is not very helpful in providing guidelines for managers because it often ignores the inevitability of inconsistencies. This leads to a proposed management strategy of 'controlled chaos' and organisational learning.

The chapter makes an important distinction between 'commodity' chemicals and speciality chemicals. This distinction is crucial to Freeman's paper on chemical innovations in Chapter 4, which also bears on strategic behaviour. He argues that the most successful innovating firms have followed a strategy of moving away from commodity to speciality chemicals and that this shift is a part of a wider change of 'techno-economic paradigm' in the economy as a whole. The evidence from empirical studies of innovation suggests that Japanese innovation management techniques and innovative strategies may enable them to overtake US and European firms, even in an industry like chemicals, in which their relative competitive strength today is less than in electronics or automobiles.

The second part of the book concentrates mainly on the *diffusion* of technical change in a service industry — banking — but also includes a chapter on the automobile industry. Banking is of particular interest because it has had the highest rate of investment in computerised equipment of any branch of manufacturing or services and yet, in most countries, this has not yet yielded the productivity gains which might have been expected. The two chapters by de Wit and by Diederen, Kemp, Muysken and de Wit throw a great deal of light on this paradox as well as making original contributions both to the theory and modelling of diffusion of technical change and to the measurement of productivity.

In Chapter 5, de Wit adapts Salter's model of technical change to take account of the specific institutional and historical features of technical innovation in the banking industry. This original development was based on thorough empirical investigation of the BGC — the Dutch automated clearing house for the commercial banks. The next chapter, by Diederen *et al.*, makes equally good use of an empirical survey of a hundred local branches which were relatively autonomous in their decision-making on when to adopt new technology; this provided a good basis for the development and testing of a novel diffusion model.

The MERIT researchers realised the need for a satisfactory diffusion model to take account of four dimensions: (1) channels of communication; (2) capacity of supply to the market; (3) skills and learning of users; and (4) organisational flexibility and change. Hitherto most diffusion models have neglected one or more of these dimensions as well as the possibilities of mismatch between them. Chapter 7, by Baba and Takai, provides an excellent demonstration of this point. It shows that Japanese management techniques have proved rather successful in achieving greater productivity gains than US banks. Whilst it is generally believed that the US software industry is ahead of the Japanese, Baba and Takai show that whether or not this is true, the Japanese methods of decentralised flexible management of innovative software adaptation to local needs have yielded superior results during diffusion of information technology in the banks. This was made possible, as in other industries, by the very close linkage between skill formation, training and localised innovation, i.e. points (3) and (4) above in the MERIT model. Thus the banking industry and the chemical industry, although differing enormously in many ways, nevertheless both illustrate some of the major conclusions emerging from Chapters 2, 3 and 4 on the strategy of firms in the management of innovation, and the importance of Japanese management techniques.

The automobile industry also is no exception. Here, too, the MIT World Vehicle Project has demonstrated the great importance of organisational innovation in achieving the large productivity advances of the Japanese car manufacturers. However, Dankbaar's chapter shows that the Japanese model of organisational change is not necessarily the only one or the best. He argues that simple imitation of Japanese methods is neither possible nor desirable, because of institutional, social and cultural differences between countries. His study of the automobile industry leads him to the conclusion that there is a West German/Swedish model of industrial organisation which is more appropriate for European conditions. He points in particular to the important role of trade unions in West Germany and Sweden in leading to a system which puts a very high premium on skill formation and workforce consultation. His paper also demonstrates well the basic advantages of the MERIT diffusion model (Chapter 6) on the positive interaction between investment, training, communication and institutional change.

From the discussion so far, it is already evident that most of the papers combine discussion of empirical research with new developments in theory, models and policy. So it would be wrong to regard Parts I and II as 'empirical' and Part III as 'theoretical'. Nevertheless it would be true to say that the five contributions grouped together in Part III do concentrate almost entirely on the development of theory and models.

Chapter 9, by Silverberg, provides a theoretical framework which embraces many of the novel developments in the theory of diffusion, both in the work of the MERIT group and other groups and individuals who have concentrated on the study of technical change. He concludes that the Schumpeterian distinction between 'innovation' and 'diffusion' can be very misleading since it obscures the numerous improvements made to a new product or process during diffusion. Furthermore, it implies that an innovation arrives as 'a consummated creation' and therefore also obscures the origins of the innovation itself. Like the chapters in Part II, Silverberg's model points to the great importance of both public and private learning, since innovations require complementary skills, tacit knowledge and all kinds of supporting structures, if they are to diffuse successfully. The decision to adopt and the calculation of the pay-back period for the potential adopter are by no means clear cut. Uncertainty still surrounds many diffusion decisions as with 'original' innovation. This leads Silverberg to reject the traditional production function approach with its assumptions of rational choice from a spectrum of well-specified alternatives.

Chapters 10 and 11 are also critical of the aggregate production function but develop alternative models which do not discard the approach altogether. Verspagen gives a lucid review of earlier work in the field before developing an original model of localised technological change following suggestions of Atkinson and Stiglitz. He then uses this model to analyse the influence of major price shocks on productivity growth.

In the following chapter van Zon, on the other hand approaches productivity growth through a vintage model of the capital stock. His framework is a 'putty-clay' model, which means that although producers may choose *ex ante* from a wide range of alternatives, once they have made their choice, labour and capital requirements are more or less fixed. This focus on *embodied* technical change in the investment in new plant and equipment is then related to the impact of R&D on productivity growth via R&D in the capital goods industry and via the internal technical activities of firms. The linking of R&D to vintage specific technological progress leads to interesting results, which as in the case of Verspagen's model may throw new light on the recent productivity slowdown.

A more microeconomic theoretical model is developed in Chapter 12 by Van Cayseele. He analyses the consequences of the inability of most innovators to follow a discriminatory pricing strategy with segmented markets for their innovations. He concludes that this problem leads to a slower rate of both process and product innovation than would otherwise be the case.

Finally, the last chapter, by Kemp and Soete, discusses one of the most fundamental contemporary problems in economic theory and one of the liveliest issues in policy debate: how to handle the true long-term costs of environmental pollution and to stimulate appropriate technologies of pollution abatement. Kemp and Soete conclude that the primary need is to shift away from *ex-post* 'clean-up' technologies increasingly towards *prevention* technologies. This, however, cannot be approached simply from the side of technology but requires an appropriate framework of economic incentives (and disincentives) which take account of the peculiar, specific features of the global pollution problem.

From this brief introduction it will be apparent that the distinction between the three parts of the book is more one of convenience than of fundamental differences. Indeed, if we accepted the full logic of Silverberg's chapter, then the distinction between 'innovation' (Part I) and 'diffusion' (Part II) would itself disappear. We hope, however, that the reader will nevertheless find the arrangement helpful in tracing the interconnections. We are ourselves only too well aware that we are trying to explore some uncharted territories and to draw some new maps. If the results are sometimes a little fuzzy and even 'chaotic' then at least in this they resemble the subject of our exploration.

# Part I: Strategy of firms and management of innovation

# 1. Inter-firm partnerships and co-operative strategies in core technologies

*John Hagedoorn and Jos Schakenraad*[1]

## 1. Introduction

In the following sections we will analyse some general patterns of inter-firm co-operation in three broad fields of technology which we perceive as core technologies. Following van Tulder and Junne (1988) and others we take new materials, biotechnology and information technologies to be the three core technologies which at present affect a large number of industrial sectors in advanced capitalist societies.[2] In this chapter we will analyse a number of aspects of technology co-operation between companies in these core technologies. This research is partly based on our CATI databank which, at present, contains worldwide information on over 7,000 co-operative agreements in a large number of technologies and several thousands of participating companies (see also Appendix I). The core of our present contribution to the analysis of inter-firm alliances is restricted to five basic issues.

We will first present some historical developments of co-operation in order to show how this phenomenon has gradually increased. This very first picture will give us some idea about the recent growth in inter-firm agreements and the differences that one finds for several fields of technology.

Secondly, we will pay attention to a number of peculiarities of different modes of co-operation. For instance, a joint venture has more effect on inter-firm co-operation than a simple licensing agreement and therefore we briefly discuss some of the major differences between a number of relevant modes of co-operation. After this we describe the distribution of these modes of co-operation for each of the core technologies.

Thirdly, we continue with a first analysis of the international distribution of technology co-operation, looking in particular at the distribution of co-operative agreements among the leading economic blocks, i.e. the USA, Western Europe and Japan.

Fourthly, we analyse the major motives that play a role in establishing co-operative agreements. These motives enable us to go one step further and examine the distribution of three basic co-operative strategies: cost motivated agreements, long-term strategic collaboration and mixed strategies, according to size of co-operating companies and core technologies.

A final exercise on this paper is the analysis of patterns of inter-firm agreements with a multidimensional scaling technique. We identify the major co-operating companies, their position in international networks and their specific linkages to other companies.

## 2. Some historical trends

General trends in the introduction of new co-operative agreements in new materials, biotechnology and information technologies, as found in our CATI databank, are presented in Figure 1. There we immediately notice the first major contrast between these fields of technology in the differences in size of populations. In our database the total number of agreements related to information technologies amounts to over 2,700 cases, which is over twice the number of agreements related to biotechnology, which counts over 1,200 cases. Biotechnology in its turn is almost twice the size of new materials, with 700 agreements. The large number of agreements in information technologies is not so surprising if we take into account that information technologies include a wide spectrum of technologies, ranging from microelectronics to telecommunications, whereas the other two core technologies still have fewer applications.

A number of trends for all these fields of technology are, however, similar. For instance, we notice the substantial growth in the number of new co-operative agreements after 1980. This trend confirms our earlier analysis (Hagedoorn and Schakenraad, 1989) and other studies which found a similar growth pattern (see, for instance, Hladik, 1985; Haklisch, 1986; Hergert and Morris, 1986; and OECD, 1986).

As can be expected, biotechnology agreements were almost non-existent before the mid-1970s when biotechnology was at a very early stage of

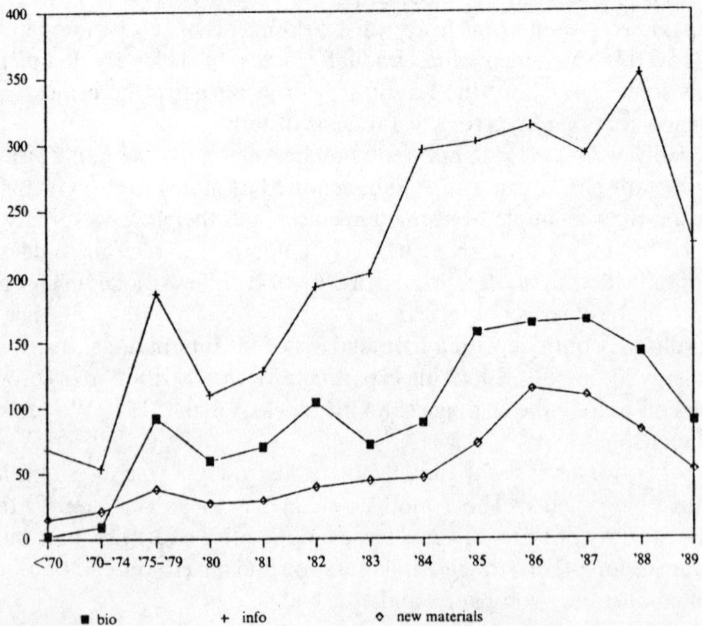


**Figure 1** Growth of newly established technology co-operation agreements in biotechnology, information technologies and new materials

*Source:* MERIT–CATI databank

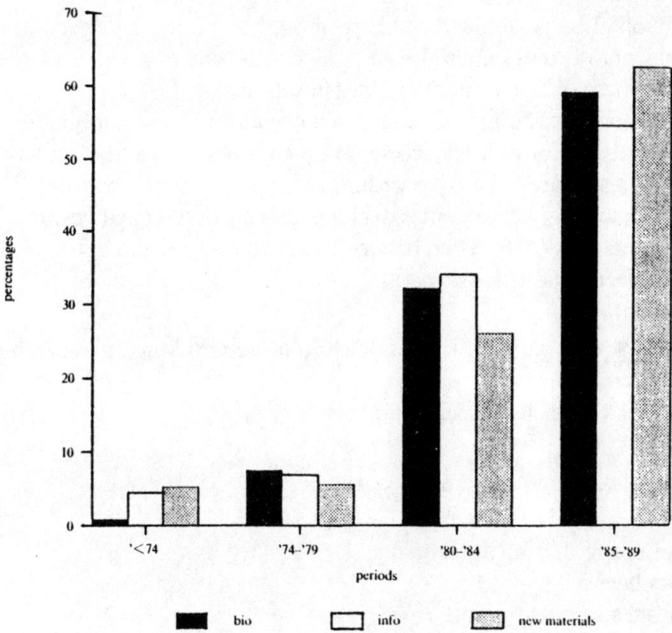


**Figure 2** Distribution of technology co-operation agreements in biotechnology, information technologies and new materials, four periods

*Source:* MERIT-CATI databank

development; for the other two fields there was already some early experience with partnering. For the first half of the 1980s we observe a steady growth of new agreements in all fields, albeit with some fluctuations. This rapid growth levelled off or declined in the mid-1980s.

From Figure 2 we learn that in all three core technologies about 90 per cent of the agreements have been established during the 1980s. In new materials over 62 per cent of the agreements were made since 1985; in biotechnology and information technologies these shares reach about 60 per cent and 54.5 per cent respectively. In other words, although technology co-operation between companies goes back many decades, it has experienced a major boost during the 1980s in these three new areas.

## 3. Organisational modes of inter-firm co-operation

Throughout our research we make a distinction between different organisational modes of co-operation. In general we distinguish joint ventures and research corporations, joint R&D agreements, technology exchange agreements, direct investment, customer–supplier relations and one-directional technology flows. All these modes of co-operation have different impacts on the character of technology sharing, the organisational context and the possible economic consequences for participating companies (see Hagedoorn, 1990). Before we comment on the distribution of these modes in biotechnology, information technologies and new materials, we will briefly mention some of the major features of these different modes of organisation.

We refer to joint ventures and research corporations as combinations of the economic interests of at least two separate companies in a 'distinct company'; profits and losses are shared according to equity investment. Joint R&D and technology exchange agreements cover agreements that regulate technology and R&D sharing and/or transfer between two or more companies. Joint R&D refers to agreements such as:

- joint research pacts which establish joint undertaking of research projects with shared resources;
- joint development agreements.

Technology exchange agreements cover:

- technology sharing agreements;
- cross-licensing;
- mutual second-sourcing.

Some equity investments can be seen as a form of co-operation between companies which in the long run could affect the technological performance of at least one 'partner'. Such minority stakes, in particular those by a large company in a smaller 'high-tech' company, can be understood as a case of co-operation, in particular if such minority sharing is coupled with research contracts.

Under customer–supplier relations we have grouped together those categories of agreements through which contract-mediated collaboration in either production or research is established. These customer–supplier relations can be divided into a number of forms of partnership:

- co-production contracts confirm the agreement between companies to produce a commodity; usually the 'leading' company supplies the technology and critical components, other companies manufacture less-critical components and assemble final products;
- 'co-makership' relations establish long-term contracts between users and suppliers with users outsourcing a part of their production process to suppliers of sub-assemblies;
- research contracts regulate R&D co-operation in which one partner, usually a large company, contracts another company, frequently a small specialised R&D firm, to perform particular research projects.

Finally, there are unilateral technology flows such as second-sourcing and licensing agreements. Advantages of second-sourcing are found in secure and overall growth of supply for one side and secure and regulated demand for the other. Licensing provides speedy entry and relatively inexpensive technology

**Table 1** Modes of technology co-operation in biotechnology, information technologies and new materials (numbers and percentages)

| | *Biotechnology* | *Information technologies* | *New materials* |
|---|---|---|---|
| Joint ventures, research corporations | 164<br>13.5% | 458<br>16.9% | 177<br>25.7% |
| Joint R&D | 362<br>29.8% | 749<br>27.6% | 173<br>25.1% |
| Technology exchange agreements | 84<br>6.9% | 328<br>12.1% | 54<br>7.8% |
| Direct investment | 234<br>19.3% | 357<br>13.1% | 65<br>9.4% |
| Customer–supplier relations | 186<br>15.3% | 245<br>9.0% | 42<br>6.1% |
| One-directional technology flows | 183<br>15.1% | 581<br>21.4% | 177<br>25.7% |
| Total | 1213<br>100.0% | 2718<br>100.0% | 688<br>100.0% |

*Source:* MERIT–CATI databank.

access to the licensee, but in general with limited sophistication of the technology.

The distribution of these modes in biotechnology, information technologies and new materials is given in Table 1. Joint R&D represents the largest group of co-operative agreements in both biotechnology and information technologies, nearly 30 per cent in biotechnology and over 27 per cent in information technologies, and also comes very close to the leading mode of co-operation in new materials, with a share of over 25 per cent. The other modes of co-operation demonstrate different patterns for each field of technology. In biotechnology direct investment is the second mode of co-operation, with over 19 per cent of all agreements, followed by customer–supplier relations and one-directional technology flows which each take over 15 per cent. This pattern is quite different from the distribution of modes of co-operation in the other core technologies and it probably reflects the particular situation in biotechnology with reference to small and large multinational companies. In biotechnology large companies are engaged in a substantial share of co-operative agreements with small, high-tech companies through minority holdings, R&D contracts and licensing agreements. Further analysis of our material, which is not represented here, demonstrates that such forms of co-operation in biotechnology are US dominated as between 30 and 50 per cent of such agreements are intra-US co-operation.

In information technologies one-directional technology flow is the second important mode with over 21 per cent of all agreements, followed by joint ventures with nearly 17 per cent, direct investment with over 13 per cent and technology exchange with over 12 per cent. The significance of one-directional technology flows is largely due to second-sourcing agreements, which are a type of co-operative agreement typical for microelectronics, and licensing which has become more important in information technologies as a number of its fields are gradually becoming more mature.

In new materials we see that there is less variation in the occurrence of different modes of co-operation as three modes constitute over 75 per cent of all agreements. Joint ventures, one-directional technology flows and joint R&D pacts each take over 25 per cent of all agreements in new materials.

## 4. International, regional distribution of co-operation

In the literature on co-operative agreements and strategic partnering the role of the so-called Triad (USA, Japan and Western Europe) is stressed by some authors to emphasise that co-operation between companies takes place especially between companies from and within these economic blocks (see, for instance, Ohmae, 1985 and the literature discussed in Hagedoorn and Schakenraad, 1989). Our present material on core technologies largely confirms our previous analyses and those made by others (Table 2). Still over 90 per cent of all agreements are made between companies from Western Europe, the USA and

**Table 2** International distribution of technology co-operation agreements in biotechnology, information technologies and new materials (numbers and percentages)

| | *Biotechnology* | *Information technologies* | *New materials* |
|---|---|---|---|
| Western Europe | 233<br>18.4% | 509<br>18.7% | 118<br>17.2% |
| Western Europe–USA | 245<br>20.2% | 599<br>22.0% | 133<br>19.3% |
| Western Europe–Japan | 38<br>3.1% | 177<br>6.5% | 49<br>7.1% |
| USA | 428<br>35.3% | 707<br>26.0% | 139<br>20.2% |
| USA–Japan | 155<br>12.8% | 406<br>14.9% | 94<br>13.7% |
| Japan | 58<br>4.8% | 95<br>3.5% | 88<br>12.8% |
| Other | 66<br>5.4% | 225<br>8.3% | 67<br>9.7% |
| Total | 1213<br>100.0% | 2718<br>100.0% | 688<br>100.0% |

*Source:* MERIT–CATI databank.

Japan.[3] There are some minor differences between core technologies. In information technologies the share of agreements with or among non-Triad country companies has risen to over 8 per cent and in new materials the share of non-Triad co-operation reaches a level of almost 10 per cent. A large section of these non-Triad agreements cover collaborative projects between companies from Triad countries with companies from South-East Asian Newly Industrialised Countries (NICs). In biotechnology about 95 per cent of the agreements are still concentrated in the Triad.

In all three fields of technology intra-US co-operation takes the largest share of all agreements, in particular in biotechnology over 35 per cent of the agreements refer to intra-US collaboration. We think that such an outcome, especially the one for biotechnology, supports the notion that the USA is still the leading region for technological development. This intra-US collaboration is followed by Western Europe–US partnerships, intra-Western European agreements and co-operation between companies from Japan and the USA.

We acknowledge that this material can present only a very first picture of the international impact of co-operative agreements. So far little is said and shown about the technology flows within these agreements, which could give a much more detailed picture of the consequences of collaboration for partners and economic blocks at large. In order to analyse such issues we shall return to that matter in future research where we will extensively apply network analysis to examine these matters in depth.

## 5. Motives for co-operation

It is obvious that companies do not co-operate for the sake of co-operation as a moral principle in a world where competition is the driving force of company behaviour. Companies have particular motives for co-operation, and these motives can be expected to differ both for the companies involved and for each individual agreement and mode of co-operation. In this section we will analyse the motives which lay behind each agreement in our databank. Details about the procedure followed to assign motives to each agreement are discussed in Appendix II. Here we will analyse motives which play a role in three major modes of co-operation: joint ventures and research corporations; joint R&D; and direct investment. Other modes such as technology exchange agreements, customer–supplier relations and one-directional technology flows will not be discussed here because these forms of co-operation are in general exclusively based on one particular motive (see Appendix II). Furthermore, from the literature we have the impression that motives for co-operation are frequently related to the modes first mentioned and not so much to the modes which are, in our opinion, more related to one particular form of co-operation such as licensing and technology sharing agreements.

In the literature we find a large number of motives which are related to co-operation between companies. In Hagedoorn and Schakenraad (1989) we have listed a number of these motives, such as:

- the extremely high costs and risks of R&D in high-tech industries;
- quick pre-emption strategies on a world scale which are preferable despite a 'loss' of potential monopoly profit;
- technology transfer and technology complementarity;
- exploration of new markets and market niches;
- shortening of period between discovery and market introduction; and
- monitoring the evolution of technologies and opportunities.

In that context the so-called 'hidden agenda' of partners is mentioned to refer to deliberate actions by some partners in joint undertakings to gain more and to share less.

In Tables 3, 4 and 5 we present the distribution of motives which we could attach respectively to: (*a*) joint venture and research corporation; (*b*) joint R&D agreement; and (*c*) direct investment arrangement in our databank.[4] Most of these motives are also mentioned in the list deduced from the literature; some are

**Table 3** Motives for joint ventures and research corporations in biotechnology, information technologies and new materials (numbers and percentages)

| | *Information technologies* | *Biotechnology* | *New materials* |
|---|---|---|---|
| Expansion/new markets | 297<br>36.8% | 75<br>22.3% | 99<br>32.9% |
| Reduction of innovation lead time | 103<br>12.7% | 62<br>18.4% | 45<br>15.0% |
| Technological complementarity | 96<br>11.9% | 63<br>18.7% | 46<br>15.3% |
| Influencing market structure | 80<br>9.9% | 24<br>7.1% | 19<br>6.3% |
| Rationalisation of production | 68<br>8.4% | 17<br>5.0% | 15<br>5.0% |
| Monitoring technological opportunities | 46<br>5.7% | 33<br>9.8% | 33<br>11.0% |
| Specific national circumstances | 36<br>4.4% | 8<br>2.4% | 6<br>2.0% |
| Basic R&D | 19<br>2.3% | 27<br>8.0% | 15<br>5.0% |
| Lack of financial resources | 17<br>2.1% | 21<br>6.2% | 5<br>1.7% |
| Other motives | 47<br>5.8% | 7<br>2.1% | 18<br>5.9% |
| Total | 809<br>100.0% | 337<br>100.0% | 301<br>100.0% |

*Source:* MERIT–CATI databank.

added and a number feature only in one or two modes of co-operation and not in the others.

It is obvious from Table 3 that only a few motives appear to play an important role in joint venture formation; they are followed by a number of other motives that have some relevance, while others which are frequently mentioned in the literature are insignificant. The search for expansion and new markets is apparently the major motive for many companies to form joint ventures. Reduction of the total period of innovation and technology complementarity are the second and third mentioned motives, but they come closer to the most frequently mentioned motive in biotechnology than in information technology and new materials. There are five other motives that deserve some attention, although their influence should not be exaggerated. These motives which play a moderate role are: influencing existing market structures through a new company; monitoring technological opportunities (in particular in biotechnology and new materials); and rationalisation of production (in particular in information technologies); in addition performing basic research and the lack of sufficient financial resources have some relevance especially in biotechnology. Other motives play only a very insignificant role; in particular the motive of reducing high costs and risks, which is so often mentioned in the literature, is only occasionally found in our data.

**Table 4** Motives for joint R&D in biotechnology, information technologies and new materials (numbers and percentages)

| | *Information technologies* | *Biotechnology* | *New materials* |
|---|---|---|---|
| Technological complementarity | 466<br>38.9% | 243<br>38.1% | 111<br>44.6% |
| Reduction of innovation time-span | 430<br>35.9% | 198<br>31.0% | 84<br>33.7% |
| Influencing market structure | 126<br>10.5% | 26<br>4.1% | 8<br>3.2% |
| Monitoring technological opportunities | 69<br>5.8% | 30<br>4.7% | 14<br>5.6% |
| Basic R&D | 41<br>3.4% | 53<br>8.3% | 21<br>8.4% |
| High costs & risks | 32<br>2.7% | 5<br>0.8% | 4<br>1.6% |
| Lack of financial resources | 31<br>2.6% | 77<br>12.1% | 7<br>2.8% |
| Other | 4<br>0.3% | 6<br>0.9% | – |
| Total | 1 199<br>100.0% | 638<br>100.0% | 249<br>100.0% |

*Source:* MERIT–CATI databank.

In Table 4 we show the distribution of motives for companies to engage in joint R&D agreements. Most of the motives listed in this table were also mentioned in the previous table, but there are some clear differences in the ranking.

As far as joint R&D projects are concerned, two motives dominate the scene. Technological complementarity and the reduction of the innovation lead time, which came second and third in joint venture formation, take a majority share of about 70 per cent or more of all motives mentioned in the three core technologies. The dominant role of these motives is due largely to their research character which can be expected to play a larger role in R&D projects than in joint ventures of which most in our databank are not exclusively research designated. Both these motives are followed at considerable distance by a small number of motives which in most cases only seem relevant to one or two fields of technology.

Influencing the market structure is apparent in information technologies which have become a relatively more mature field with a more or less mature market structure. This motive appears to be of less relevance in biotechnology and new materials. Lack of financial resources appears to have some influence only in biotechnology; the role of this motive in information technologies and new materials is very small. We think that this reflects the special situation in biotechnology, where a substantial number of sophisticated small and sometimes medium-sized companies performed expensive research, but overshot their financial capabilities and, consequently, were forced to join with another company.

Monitoring technological opportunities and performing basic research have some significance in biotechnology and new materials but little in information technologies. For both these motives this contrast is presumably due also to differences in technological maturity; biotechnology and a number of activities in new materials still being in an earlier phase of many of their technological trajectories where co-operation in basic research pays off.

As with joint ventures, we notice that many motives mentioned in the literature are not as important as first thought, or, as in the case of high costs and risks, appear to be almost irrelevant.

Table 5 shows the motives for direct investment as a mode of co-operation. Even more than in the two foregoing tables, we notice that there are a number of differences between the ranking of motives in biotechnology, information technologies and new materials. In biotechnology and information technologies the technological competence of the partner is listed as the most important motive, but this motive is almost twice as important in biotechnology as in information technology. In information technology there is an almost even distribution of most motives, indicating that most of them have a fair significance, with the exception of downstream entry, which is of little importance.

In biotechnology the technological competence of partners is a major motive for direct investment deals, as it takes a share of over 35 per cent. It is followed by monitoring of possible entry and the profitability of partners as two other important motives. Control over partners and tightening of customer–supplier

**Table 5** Motives for direct investments in biotechnology, information technologies and new materials (numbers and percentages)

| | *Information technologies* | *Biotechnology* | *New materials* |
|---|---|---|---|
| Technological competence of partner | 103<br>18.9% | 127<br>36.7% | 16<br>17.4% |
| Monitoring of possible entry | 78<br>14.3% | 73<br>21.1% | 19<br>20.7% |
| Profitability of partner | 75<br>13.8% | 62<br>17.9% | 5<br>5.4% |
| Control over partner | 97<br>17.8% | 30<br>8.7% | 5<br>5.4% |
| Tighten c-s relation | 82<br>15.0% | 32<br>9.2% | 18<br>19.6% |
| Horizontal matching of core activities | 86<br>15.8% | 20<br>5.8% | 24<br>26.1% |
| Downstream entry | 24<br>4.4% | 2<br>0.6% | 5<br>5.4% |
| Total | 545<br>100.0% | 346<br>100.0% | 92<br>100.0? |

*Source:* MERIT–CATI databank.

relationships are relevant, but clearly less important than in information technologies. Horizontal matching of core activities, which does play a role in information technologies and particularly in new materials, is of only limited relevance in biotechnology.

We have already mentioned that, compared to the other core technologies, direct investment plays only a limited role as a mode for inter-company co-operation in new materials. Horizontal matching of core activities accounts for over one-quarter of all motives. It is followed by two motives each with a share of about 20 per cent, i.e. monitoring of possible entry and the wish to tighten customer–supplier relationships. The fifth motive which plays some role is the technological competence of partners, with a share of about 17 per cent. It is clear that the other motives listed in Table 5 play only a role of minor importance in new materials.

So far we have seen that only a relatively small number of motives matter for co-operation in these core technologies. Motives have a different bearing for different modes of co-operation, but in general the search for new markets and entry, the reduction of period of innovation, the technological complementarity of partners and monitoring technological opportunities are the major motives we have come across. Sharing of costs and risks which are often associated with inter-firm collaboration apparently play a negligible role.

## 6. Strategic backgrounds in inter-firm co-operation

In a number of contributions to the study of industrial organisation, so-called intermediate or mixed forms of governance between markets and integration have been theorised (see especially Williamson, 1985; Riordan and Williamson, 1985). Most of the modes of co-operation we distinguish in our research fall within this intermediary level of governance. We discuss some of the broader consequences of transaction cost theory, and other relevant economic theories for this field of research, elsewhere (see Hagedoorn, 1989c). Here it will suffice briefly to mention some of its complications for the empirical analysis of technology co-operation.

A major shortcoming of the transaction cost contribution in the analysis of mixed modes of economic organisation is its narrow focus on costs of economic exchange. These mixed modes, such as joint ventures and other joint activities, are analysed merely in terms of the costs related to the so-called 'make or buy' decision. Furthermore, vertical relationships between companies receive more attention than those which have a primarily lateral character. Consequently, the (transaction) cost aspect of mixed modes is overstated and the long-term perspective of strategic impacts on both vertical and horizontal relationships between companies is largely neglected.

In a number of other contributions to the analysis of technology co-operation the strategic impact is stressed, with less attention to cost economising. In our opinion both the strategic long-term positioning of companies with respect to product-market combinations and their cost sharing behaviour, and possible combinations of these two options should be taken into account when the wide range of intermediate modes of economic organisation is studied in detail.

In this section we will first make an attempt to 'measure' the long-term strategic character or the cost economising qualities for each agreement in one of the core technologies in our databank. In short, every agreement is evaluated as having:

- mainly long-term strategic implications; or
- whether it is in particular associated with the control of (transaction) costs; or
- in case both general motives seem possible, as being of a mixed character.[5]

From Figure 3 we learn some differences between the three core technologies in general strategies towards co-operation. It is clear that in all three core technologies long-term strategic positioning is the major objective of nearly half or over half of the agreements. In biotechnology more than 50 per cent of the agreements are related to long-term positioning and only one-quarter of the agreements are aimed at cost-economising effects. In both information technologies and new materials the strategic content of agreements is somewhat less dominant than in biotechnology and cost economising is clearly more important.

In order to obtain some insight into the historical development of strategic objectives in each core technology, we present an overview of the distribution of

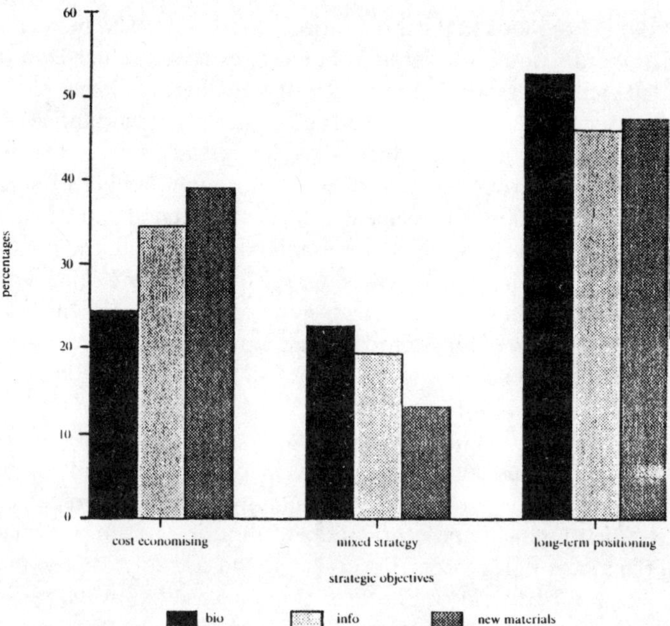


**Figure 3** Distribution of strategic objectives of co-operative agreements in biotechnology, information technologies and new materials

*Source:* MERIT–CAT databank

strategic objectives in three periods for each core technology in Tables 6, 7 and 8.

The distribution of strategic objectives for co-operation in biotechnology is given in Table 6. We had already noticed in Figure 2 that most of the agreements in biotechnology, that is about 60 per cent, were established in the second half of the 1980s.

**Table 6** Strategic objectives for co-operation in biotechnology (numbers and percentages)

| | *Before 1980* | *1980–84* | *1985–89* | *Total* |
|---|---|---|---|---|
| Cost economising | 12 | 89 | 196 | 297 |
| | 11.7% | 22.7% | 27.3% | 24.5% |
| Mixed strategy | 23 | 89 | 164 | 276 |
| | 22.3% | 22.7% | 22.8% | 22.8% |
| Long-term positioning | 68 | 214 | 358 | 640 |
| | 66.0% | 54.6% | 49.9% | 52.8% |
| Total | 103 | 392 | 718 | 1213 |
| | 100.0% | 100.0% | 100.0% | 100.0% |
| | 8.5% | 32.3% | 59.2% | 100.0% |

*Source:* MERIT–CATI databank.

If we take a closer look at the three periods, it becomes clear that a number of changes in the distribution of strategic objectives have occurred. In the years before 1980, when there were still very few agreements in biotechnology, 66 per cent of these agreements were identified as long-term strategic and very few had cost economising as a central focus. As we already know, the 1980s are characterised by a substantial growth in the absolute number of agreements. However, the proportion of agreements with cost reduction and/or cost sharing as the major objective grew even more rapidly. Comparing the period before 1980 with the second half of the 1980s, we see that the share of cost-motivated agreements has risen from nearly 12 per cent to over 27 per cent. In the same two periods the proportion of long-term strategic agreements has decreased from 66 per cent to about 50 per cent. The major cause of this development is to be found in changes that occurred in biotechnology research within companies. In the period before 1980 biotechnology research was still in its infancy years and costs, although by no means completely irrelevant, were not the major matter of concern. At present a large share of biotechnology research in companies reaches the stage of development and near-commercialisation. It is in particular in this phase of the innovation process that costs rise considerably. It is estimated that in general over 80 per cent of the costs of innovation fall in the post-research phase (see for instance Kay, 1979, pp. 26, 27 and 233). This cost pattern would explain why the category of cost-motivated agreements has increased in the past years and why lack of financial resources is a motive for some biotechnology firms (see Table 4) to engage in co-operative agreements.

From what we have already learned about differences between biotechnology and information technologies, it will not come as a surprise that the patterns of strategic motives in these fields of technology differ considerably. Table 7 shows that in information technologies cost-motivated objectives are even more important than in biotechnology. However, in recent years the number of cost-motivated agreements shows a slight decline from about 37 per cent before the mid-1980s to 32.5 per cent in the second half of the decade. Mixed strategies, in

**Table 7** Strategic objectives for co-operation in information technologies (numbers and percentages)

| | *Before 1980* | *1980–84* | *1985–89* | *Total* |
|---|---|---|---|---|
| Cost economising | 113<br>36.6% | 345<br>37.2% | 481<br>32.5% | 939<br>34.5% |
| Mixed strategy | 48<br>15.5% | 192<br>20.7% | 290<br>19.6% | 530<br>19.5% |
| Long-term positioning | 148<br>47.9% | 390<br>42.1% | 711<br>48.0% | 1 249<br>46.0% |
| Total | 309<br>100.0%<br>11.4% | 927<br>100.0%<br>34.1% | 1 482<br>100.0%<br>54.5% | 2 718<br>100.0%<br>100.0% |

*Source:* MERIT–CATI databank.

which both costs and long-term positioning play a role, stabilised at the level of about 20 per cent throughout the 1980s, after reaching only about 15 per cent in the period before 1980. The share of 'real' strategic alliances amounted to 48 per cent in the period before 1980, then decreased to 42 per cent in the first half of the 1980s and regained a share of 48 per cent in the final period.

The explanation for this particular pattern could be related to some general trends in information technology. In the period before 1980 a number of fields in information technology were still in a relatively early period of establishing a technological paradigm, whereas others had already reached a more mature level of development. This explains why both cost- and strategically-motivated agreements take such a large share of all agreements in those years, a picture which is in sharp contrast to biotechnology in the same period. In the first half of the 1980s some fields of information technology entered into a somewhat more mature phase, and mixed agreements with both an element of cost reduction and strategic positioning became more important. Then, in the second half of the 1980s, the picture of co-operation in information technologies changed. Although the number of cost-motivated agreements rose in absolute terms, its relative position decreased on account of strategic alliances which recaptured their early share. We expect that most of these new strategic alliances have to be seen in the light of three important developments in industrial restructuring which is at present taking place in information technologies. One development is the absolute and relative growth of agreements in software and telecommunications in which strategically motivated alliances are dominant.[6] The second development is the strategic restructuring which takes place within several fields among international and diversified companies which form strategic alliances within particular fields of information technology, such as alliances within microelectronics and telecommunications. The third development can be characterised as the creation of inter-sectoral relationships in which possible technological and business linkages between several fields of informa-

**Table 8** Strategic objectives for co-operation in new materials (numbers and percentages)

| | *Before 1980* | *1980–84* | *1985–89* | *Total* |
|---|---|---|---|---|
| Cost economising | 17<br>22.9% | 94<br>50.8% | 158<br>36.8% | 269<br>39.1% |
| Mixed strategy | 6<br>8.1% | 17<br>9.2% | 69<br>16.1% | 92<br>13.4% |
| Long-term positioning | 51<br>69.0% | 74<br>40.0% | 202<br>47.1% | 327<br>47.5% |
| Total | 74<br>100.0%<br>10.7% | 185<br>100.0%<br>26.9% | 429<br>100.0%<br>62.4% | 688<br>100.0%<br>100.0% |

*Source:* MERIT–CATI databank.

tion technology, such as computers and telecommunications, or software and all the other fields, are being pursued.

The distribution of strategic objectives for co-operation in new materials is presented in Table 8. First of all, we notice that for the whole period nearly half the agreements in new materials can be characterised as long-term, strategic positioning. However, also close to 40 per cent of the agreements can be labelled as cost-economising arrangements. Before 1980 almost 70 per cent of the agreements had a strategic content and less than one-quarter of the agreements were aimed at cost economising. This changed considerably during the first half of the 1980s, when over 50 per cent of the agreements were directed towards cost economising and the share of long-term strategic partnering dropped from nearly 70 per cent to 40 per cent, and cost-motivated co-operation more than doubled from 23 per cent to 50 per cent of all agreements in new materials. In the second half of the 1980s the picture changed again: strategically motivated agreements climbed to 47 per cent, the share of cost-economising motivated agreements dropped from 50 per cent to 37 per cent and the share of mixed strategies rose to 16 per cent. Although not represented here, disaggregated analysis of our material suggests that the growth of strategic alliances, in particular sub-fields such as technical ceramics and electronic, magnetic and optical materials, causes this recent relative increase of long-term positioning alliances in new materials.

More detailed analysis by size of company shows that in biotechnology and new materials larger companies tend to follow a slightly more strategic co-operative behaviour than smaller companies. In information technologies this is not so clear, but there a larger proportion of small companies follow a cost-economising behaviour in their co-operative strategy.

## 7. The structure of strategic partnering

In order to analyse the degree of co-operation between companies in each core technology we applied a non-metric multidimensional scaling (MDS) technique which we have already introduced in previous analyses (see Hagedoorn and Schakenraad, 1989). MDS is a data reduction procedure comparable to principal component analysis and other factor analytical methods. One of the main advantages of MDS is that usually, but not necessarily, MDS can fit an appropriate model in fewer dimensions than can factor analytical methods. MDS algorithms offer scaling of a similarity or dissimilarity matrix into points lying in an X-dimensional space. The purpose of this method is to provide co-ordinates for these points in such a way that distances between pairs of points fit as closely as possible to the observed (dis)similarities. In order to facilitate interpretation, the solution is given in two dimensions, provided that the fit of the model is acceptable. A stress-value indicates the goodness-of-fit of the configuration.

The total number of strategic partnerships between two companies (excluding all cost-economising motivated agreements) is taken as a measure of similarity between those two companies. A large similarity, as found in a similarity matrix, indicates intensive co-operation.[7] Unfortunately our MDS software can only analyse (dis)similarity matrices to a maximum of 45 rows, which means that for each core technology the analysis is restricted to a maximum of 45 companies with the largest numbers of strategic partnerships. For an explanation of company codes we refer to Appendix III.

For each core technology MDS solutions are presented for two periods: the years from 1980 to 1984, and from 1985 to 1990. For the first period (1980–84) we have taken those alliances established in that particular period plus those alliances made before 1980 that were not already discontinued in 1980. For the second period, the years since 1985, we follow the same procedure; we have taken all alliances forged in that period and added those alliances from the earlier

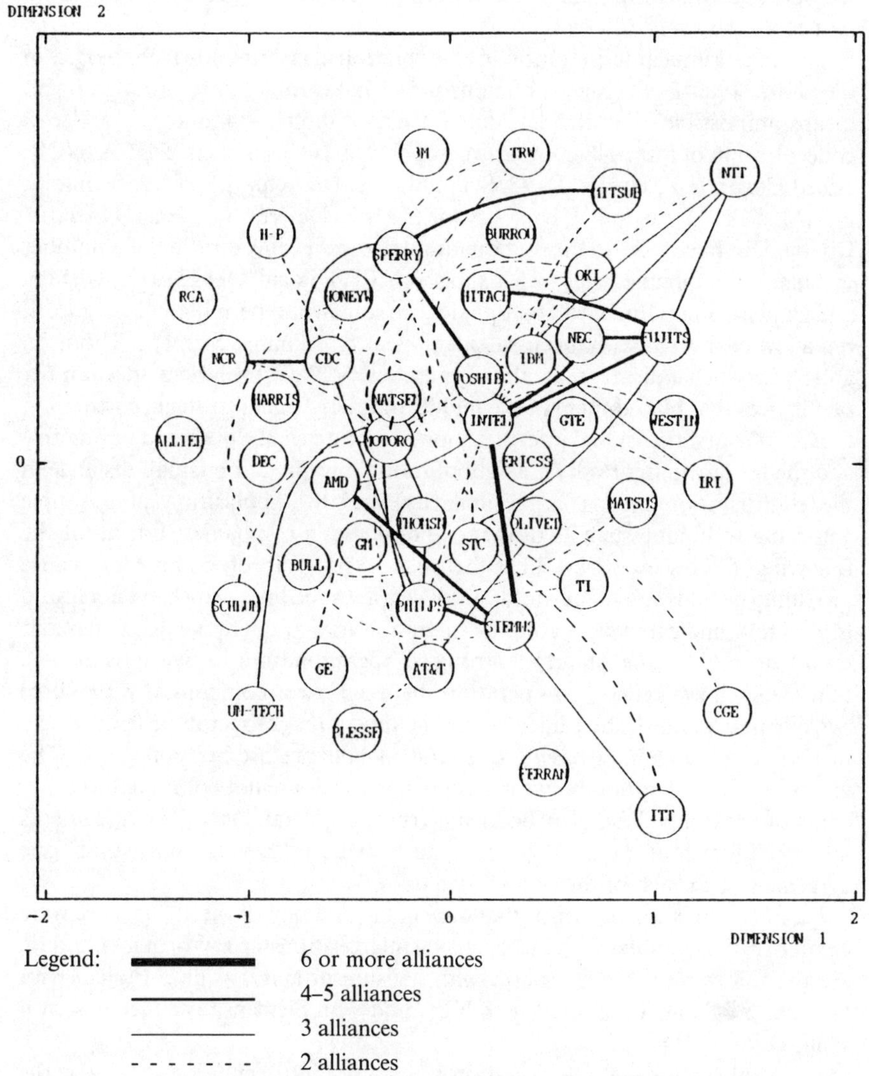


Legend: 6 or more alliances; 4–5 alliances; 3 alliances; 2 alliances

**Figure 4** The structure of strategic partnering in information technologies, 1980–84

period which were not already discontinued in 1985. During each period alliances of subsidiaries and divisions are assigned to the parent company. Also within each period, the existing alliances of companies taken over by others or partnerships made by merging companies will be assigned to the acquiring company or the new corporation.

For information technologies the intensity and structure of co-operation in two periods is pictured in Figures 4 and 5, based on the results of the non-metric scaling procedure described above. The stress-values for both periods amount to 0.115 and 0.134, which are generally seen as a fair goodness-of-fit value. Interpretation of both figures can take place in two ways. Dimensional interpretation is the most common approach used in MDS, as well as in factor methods. The first dimension is the most important as it accounts for the greatest part of the observed (dis)similarities.

The dimensional interpretation of co-operation in information technologies in the period 1980–84 (Figure 4) remains somewhat troublesome, although by no means impossible. On the left hand side of dimension one we notice a concentration of intra-US company co-operation by firms such as RCA, NCR, Allied Corp., H-P, Harris, DEC, Schlumberger (Fairchild), United Technologies (Mostek), Honeywell, CDC, Advanced Micro Devices and General Electric. On the right hand side, in the first quadrant, we see a concentration of a number of Japanese companies such as Mitsubishi, NTT, Hitachi, Oki Electric, Toshiba and Fujitsu. Some European companies are somewhat 'peripheral' in Figure 4, but a few have a somewhat central position such as Philips, Siemens, Thomson and STC, which indicates that, already in the early 1980s, the latter had a number of alliances that placed them in the middle of international strategic partnering.

Apart from dimensional interpretation, structure can be observed in Figure 4 and the following figures by a neighbourhood interpretation (small distance in the configuration means large similarity) and the application of a simple clustering technique such as drawing lines between companies. For all of the following figures we drew lines between every pair of companies whose proximity exceeds some threshold value. Very fat solid lines, which are not found in Figure 4, indicate very strong co-operation (six co-operative agreements or more), normal fat lines stand for strong co-operation (four or five agreements), thin solid lines reflect co-operation between two companies with three agreements, while dashed lines represent moderate co-operation (two agreements). Two companies having one strategic alliance are not connected. The intensity of co-operation between companies which are not connected to other companies through lines is by no means truly peripheral. Only their agreements are spread over a number of companies without having more than one agreement with each of the other companies.

From Figure 4 we also learn that during the first half of the 1980s there were a number of companies with rather strong interrelationships. Worth mentioning are the following couples: Sperry with Mitsubishi and Toshiba, Fujitsu with Hitachi, NEC and Intel, Intel with NEC and with Siemens, and Siemens with Philips and AMD.

If we take a look at the situation in information technologies during the second half of the 1980s, in Figure 5, we notice that there is still a concentration

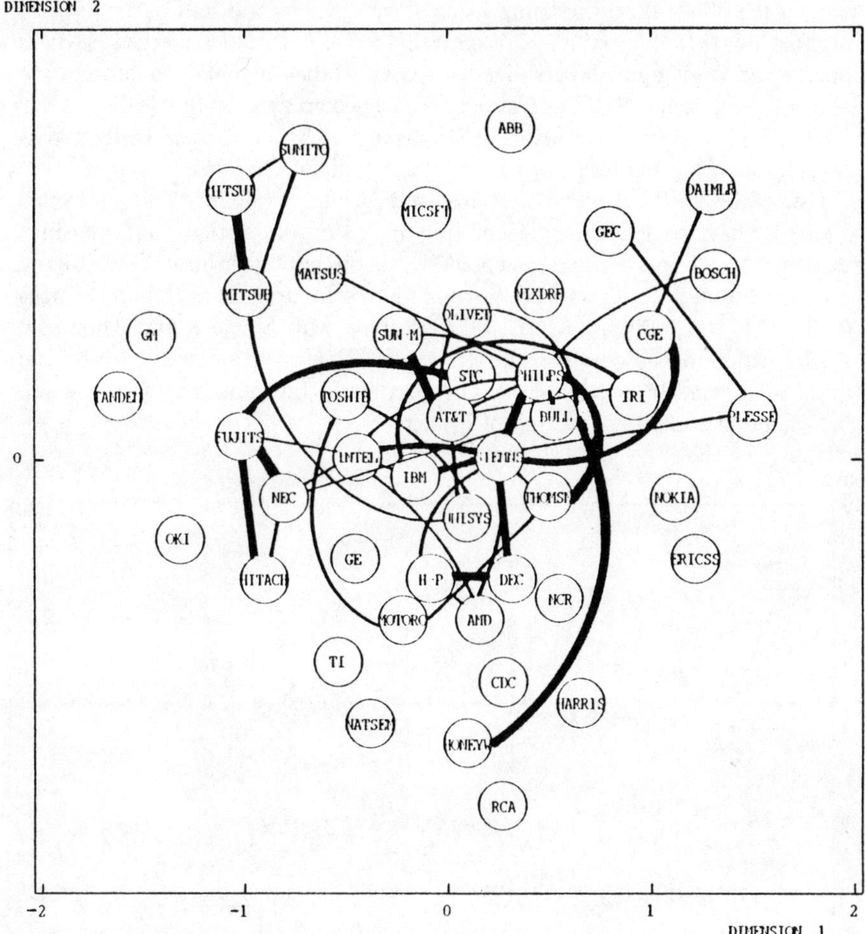


Legend: see Figure 4

**Figure 5** The structure of strategic partnering in information technologies, 1985–89

of intra-Japanese strategic partnering with almost identical companies as during the previous period. On the right hand side of dimension one we see a concentration of intra-European co-operation in particular by companies such as GEC, CGE, Daimler-Benz (through its acquisition of AEG), Bosch, IRI, Plessey, Nokia and Ericsson. In the centre of both dimensions we find the companies that have most international strategic partnerships with a number of companies. Here we find a number of the world leading companies in information technologies such as IBM, Unisys, Siemens, ATT, Thomson, Bull, Philips, STC, GE, Toshiba, Intel, H-P and DEC. Compared to the previous period we notice that most of the largest US and European companies, led by Siemens and IBM, have moved towards the centre, which indicates their dominant role in worldwide partnering.

In the previous period there were a substantial number of US companies among the 45 most co-operating firms; during the second half of the 1980s a number of these companies 'disappeared'. Some companies merged, such as Sperry and Burroughs, others divested (parts of) their activities in information technologies, such as ITT's divestment of its telecom division to Alcatel (CGE), United Technologies sold Mostek to Thomson, and Schlumberger finally sold its troublesome chip manufacturer Fairchild to Fujitsu.

One of the consequences of the substantial growth of alliances in recent years is that we have to delete the presentation of clusters of less than four inter-firm agreements, otherwise Figure 5 would be one big jumble of lines. From Figure 5 we learn that a number of very strong tie-ups are apparent, such as Siemens with IBM, Intel, Philips, CGE, DEC; Philips with Siemens and Thomson; AT&T with Sun Microsystems; Fujitsu with STC, Hitachi and NEC, Mitsubishi and Mitsui, and, before Honeywell divested its IT business, a very strong link between that company and Bull of France.

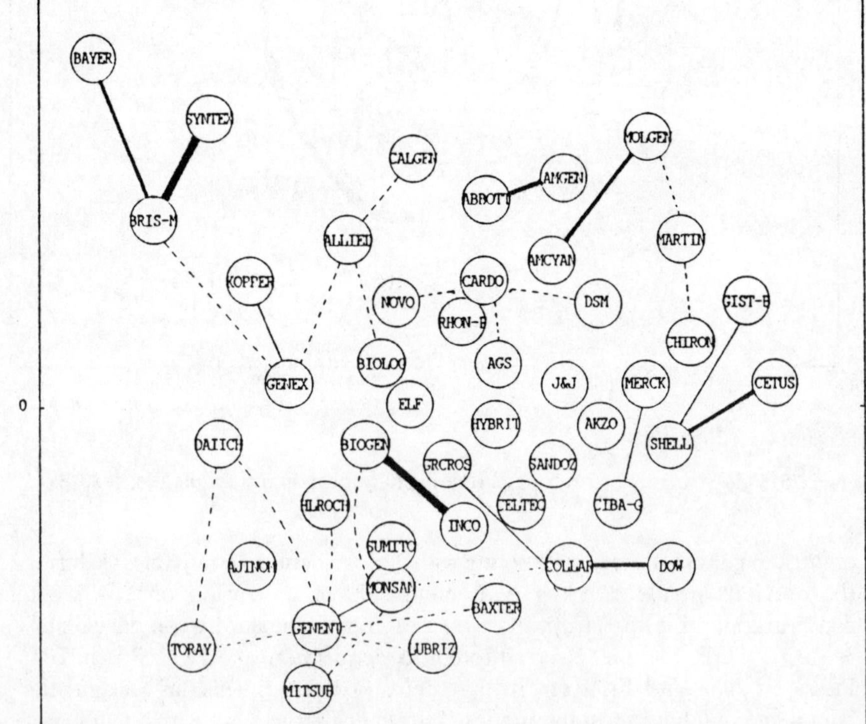


Legend: see Figure 4

**Figure 6** The structure of strategic partnering in biotechnology, 1980–84

In Figures 6 and 7 we present the MDS procedure for biotechnology; the stress-values for both periods are 0.046 and 0.071, which have to be seen as a good fit. In the period 1980–84 we see that there is hardly a pattern to be discovered. Still a number of features are worth mentioning. First of all it is clear that US companies play an important role in strategic alliances in biotechnology. About 50 per cent of the most co-operative companies are US companies, not only in this period but also during the second half of the 1980s. In the first half of the 1980s we also see that a number of 'young' companies such as Biogen, Genex, Genentec and Collaborative Research have a number of partnerships with large companies.

It seems that there is a concentration of intra-Japanese co-operation of companies such as Dai-ichi, Ajinomoto, Toray, Mitsubishi and Sumitomo with a number of other companies such as Genentech, Monsanto and Hoffmann-La Roche at the left hand side of dimension one in Figure 6. As far as most

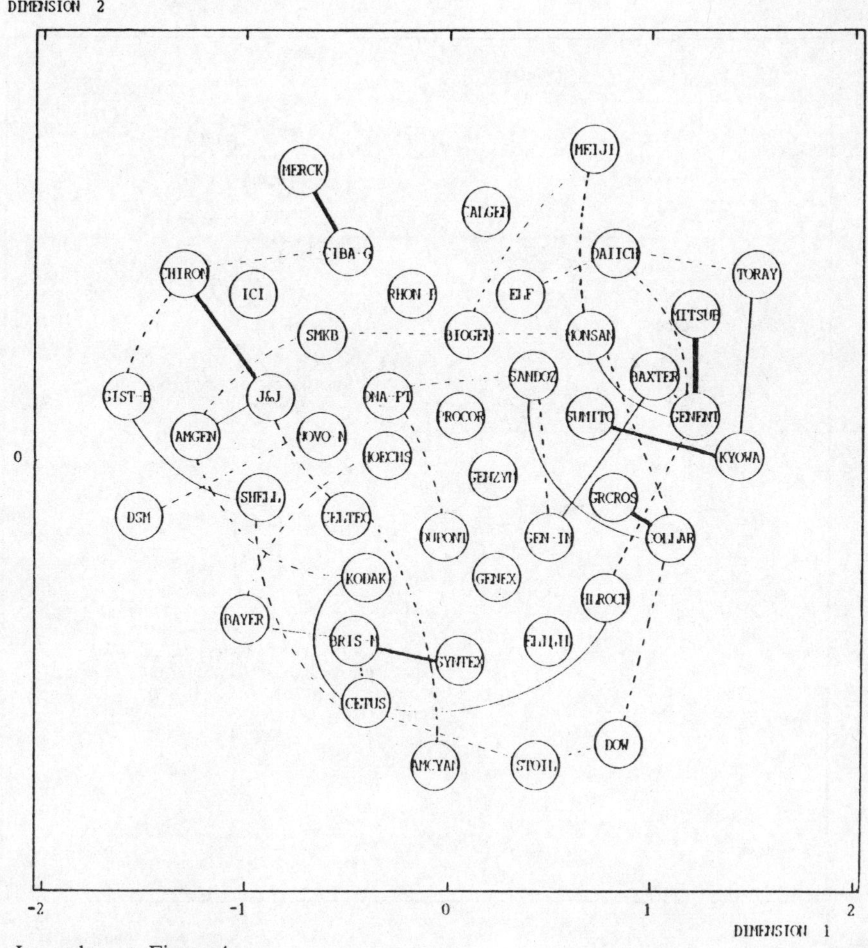


Legend: see Figure 4

**Figure 7** The structure of strategic partnering in biotechnology, 1985–89

European and US companies are concerned, there is no clear pattern in terms of a concentration in economic blocks.

Compared to information technologies we see that there is only a small number of intensely co-operating couples of companies. Only Biogen and Inco, and Syntex and Bristol-Myers are strongly interconnected, followed by partnerships such as Bayer–Bristol Meyers, Abbott Laboratories–AmGen, Molecular Genetics–American Cyanamid, Collaborative Research–Dow, and Shell and Cetus.

As shown in Figure 7 this structure of inter-firm alliances in biotechnology did not really change during the second half of the 1980s. Intra-Japanese co-operation is concentrated within a small group of companies. Most of these companies have been mentioned above, with the exception of Green Cross, which entered into closer Japanese co-operation, and Kyowa Hakko Kogyo, which was not represented in the top 45 of co-operating companies in

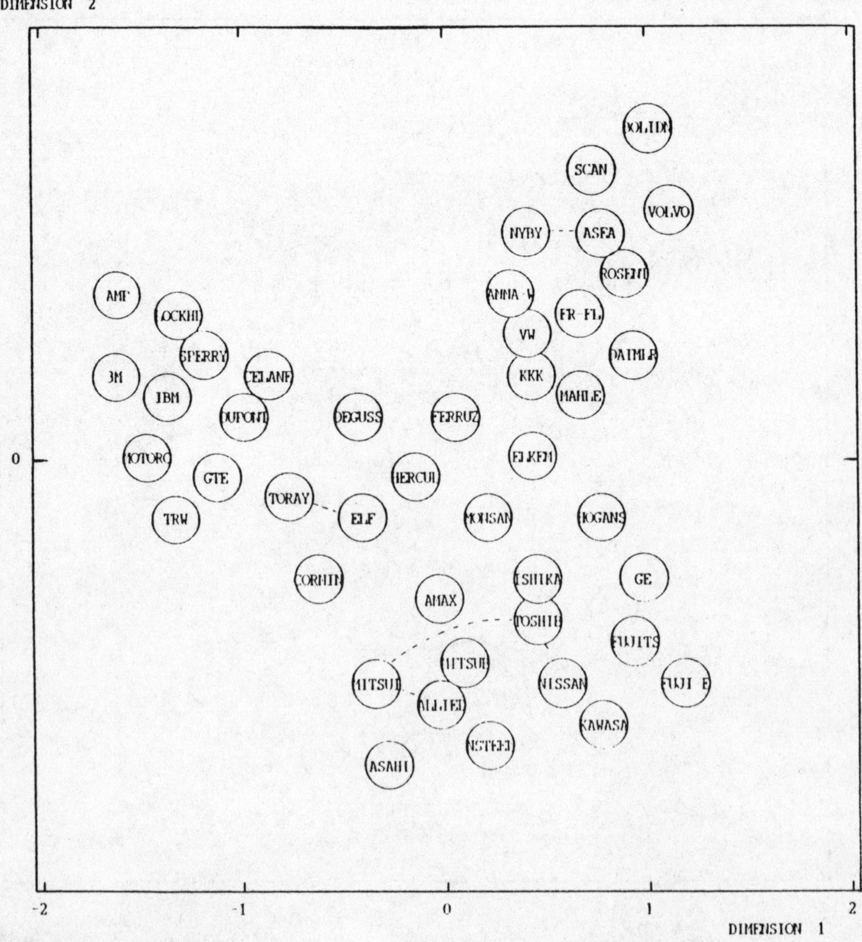


Legend: see Figure 4

**Figure 8** The structure of strategic partnering in new materials, 1980–84

biotechnology in the previous period. Most of the European companies are found on the left-hand side of dimension one, suggesting a degree of similarity in their pattern of co-operation, whereas US companies are dispersed throughout Figure 7.

As with the period 1980–84, we see only a small number of strong tie-ups in biotechnology, of which none is identical to the previous period. In the second half of the 1980s these strong partnerships existed between Merck and Ciba-Geigy, Chiron and Johnson & Johnson, Mitsubishi and Genentech, Sumitomo and Kyowa, Green Cross and Collaborative Research, and Bristol-Myers and Syntex.

Both the MDS solutions for new materials are given in Figures 8 and 9, where the goodness-of-fit for the first period is perfect, with a value of 0.014, and good for the second period, with a value of 0.050. As expected from our discussion in the previous section, we notice that the intensity of inter-firm alliances in new

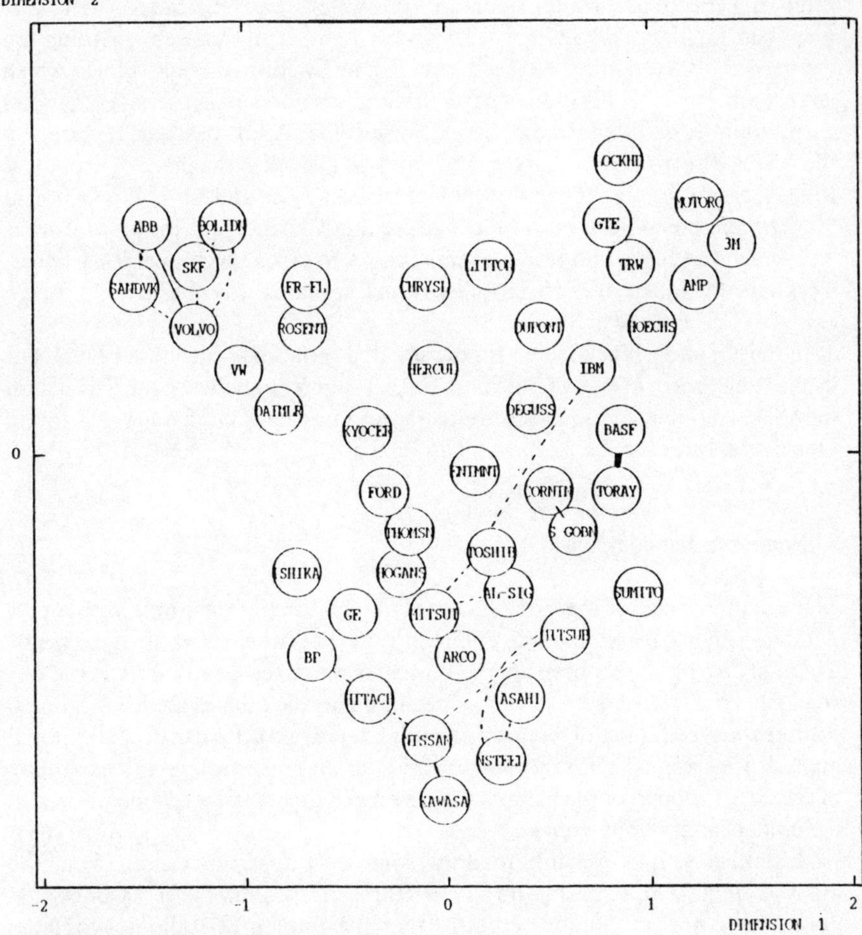


Legend: see Figure 4

**Figure 9** The structure of strategic partnering in new materials, 1985–89

materials is rather moderate because most of the inter-firm partnerships are single agreements.

For the period 1980–84 we can discover three main blocks of partnerships in new materials. On the left-hand side of dimension one in Figure 8 we notice the close similarity of a number of US companies such as AMP, Lockheed, Sperry, Celanese, 3M, IBM, Dupont, Motorola, GTE and TRW. On the other side of Figure 8 we see a concentration of both European and largely Japanese co-operation. In the European block we can identify a distinct Swedish sub-block with companies such as Boliden, Scandust, Volvo, ASEA and Nyby Uddeholm Powder; and a West German sub-block with Anna Werke, Rosenthal, VW, Friedrich Flick, Daimler-Benz, KKK and Mahle. In the Japanese block we find companies such as Ishika-Wajima Harima, Toshiba, Fujitsu, Nissan, Kawasaki, Nippon Steel, Asashi Glass and Mitsubishi.

In the second half of the 1980s this pattern of alliances in new materials was largely identical with some minor changes such as the growth of the number of multi-agreement partnerships (Figure 9). Within the three blocks that were identified for the previous period, we see a number of changes regarding the companies involved in strategic alliances. In the Swedish–German block, which largely represents the European participation, we see a smaller set of companies with a number of different companies. For instance ASEA has been replaced by the ASEA-Brown Boveri merger, and Sandvik has taken the place of Scandust. In the German sub-block KKK, Mahle, Friedrich Flick and Anna Werke are no longer represented. In the US block we see that Chrysler and Litton Industries have joned this block and Hercules came closer to other US companies, whereas Celanese was taken over by Hoechst, which explains why the latter is drawn closer to US companies.

In the Japanese block Ishika is now situated somewhat outside the block but its place has been taken by Hitachi. A further difference between the first half of the 1980s and the second half is that a larger number of companies was found outside the three blocks we have mentioned above.

## 8. Some concluding remarks

This chapter is one of the first outcomes of a research programme with a new database on partnerships and companies participating in such agreements. Therefore we have only been able to present some of the very first results of our research on a restricted number of technologies. We have merely begun to explore the possibilities of research on our material and a good deal of additional analysis is necessary before we might arrive at a more thorough understanding of the larger context of the major subjects related to inter-firm co-operation and technological development.

Nevertheless, it is possible to draw some very first conclusions from the analysis of inter-firm partnering in the three core technologies as presented above. Our present findings confirm that the forging of technology-related partnerships is to a very large extent restricted to companies from and within the Triad USA, Japan and Europe. Over 90 per cent of the agreements we found are related to companies from these economic blocks. Although there are differences between the fields of technology discussed in the above, and in recent years a larger share of companies from some NICs do seem to participate in such partnerships; by and large technology co-operation appears to remain a game of the USA, Japan and Western Europe.

It is also clear that large-scale technology co-operation in the three technologies is a relatively young phenomenon which developed particularly during the first half of the 1980s and then levelled off somewhat. This does not necessarily suggest that partnering between companies will disappear in the short run, but more probably a certain degree of stabilisation will follow the sharp increase of the past period. We have also seen that this growth in the number of agreements in all three core technologies takes place within a wider array of different modes of co-operation.

Two important issues with respect to inter-firm technology agreements are the motives which lead companies to engage in such partnerships and their strategic behaviour regarding co-operation. From the literature one learns that a large number of motives could lead companies to forge partnerships. So far we found that only four motives appear to play a major role, i.e. the possibility to find and enter new markets, the reduction of the period of innovation, the technological competence of partners, and the prospect of monitoring technological opportunities. Financial constraints, which are frequently referred to in the literature, appear to be relatively unimportant, with the exception of biotechnology where the lack of financial resources seems to have played some role. In order to determine the strategic background of co-operation we have made a distinction between cost-economising behaviour, long-term positioning and a mixed alternative which involves both options. The analysis of agreements as well as the analysis of companies revealed that long-term positioning is the major objective of co-operation in each of the three core technologies. Long-term positioning has been somewhat more emphasised in the co-operative strategy of companies in biotechnology than in the other two core technologies, although cost-economising strategies have recently also become more important in the former field. In information technologies and new materials the relevance of long-term positioning strategies appear to have increased again in recent years. We have offered some preliminary explanations for a number of differences between the core technologies and some historical changes in the pattern of cooperative strategies but further analysis of the developments within each of these core technologies is necessary to achieve a better understanding of these patterns and changes.

If we consider the role that different size classes of companies play, we have to interpret our findings with care because the differences are relatively small. It seems that, as expected, larger companies are more often following a long-term positioning strategy than smaller companies which tend to demonstrate a somewhat more cost economising behaviour, but these differences are not truly significant.

Finally, our analysis of the structure of strategic partnerships in each core technology has demonstrated that in each of the core technologies the world leading companies play a very active role. In information technologies we see that the networks of companies have become more dense since the second half of this decade. There are a large number of strong partnerships between many of the leading companies, which are not only visible in information technologies but also in biotechnology and in new materials. In biotechnology US companies play an important role in forming alliances as they are well represented among the most collaborative companies. Japanese companies seem to remain largely within their own economic block. In new materials the emergence of three blocks of companies is most noticeable. There we see three clearly different blocks of co-operative companies in a US, Japanese and European (i.e. German–Swedish) formation. In other words, globalisation of strategic alliances between the world-leading companies exists in all three core technologies, but so far it appears most noticeable in information technologies, i.e. the most mature field of these core technologies.

## Appendix I The Co-operative Agreements and Technology Indicators (CATI) information system

The CATI databank is a relational database which contains separate data files that can be linked to each other and provide (dis)aggregated and combined information from several files. So far information on over 7,000 co-operative agreements has been collected. Major sources for our databank are newspaper articles, books dealing with the subject, and in particular specialised technical journals which report on business events as well. This method of information gathering has its drawbacks and limitations:

- in general only those agreements which companies make public are published;
- newspaper and journal reports are likely to be incomplete, especially when they go back in history and/or regard firms from countries lying outside the scope of the journal;
- a low profile of small firms without well-established names is likely to have them excluded; and
- another problem is that information about the dissolution of agreements is generally not published.

Despite these shortcomings, we think we have been able to produce a databank which enables us to perform empirical research which goes beyond case-studies or general statements.

Our databank contains information on each agreement and some information on companies participating in these agreements. First of all, we defined co-operative agreements as common interests between independent partners which are not connected through (majority) ownership. In the CATI information system only those inter-firm agreements are being collected that contain some arrangements for transferring technology or research. Joint research pacts, second-sourcing and licensing agreements and research corporations are clear-cut examples. We also collect information on joint ventures in which new technology is received from at least one of the partners, or joint ventures having some R&D programme. Mere production or marketing joint ventures are excluded.

We regard as a relevant input of information for each alliance:

- the number of companies involved;
- names of companies (or important subsidiaries) in the agreement;
- involvement of banks, research institutes, states, universities and venture groups;
- time-horizon of the agreement;
- year of establishment;
- duration of the alliance;
- capital investment and potential additional investments;
- field(s) of technology, such as:[8]
  - biotechnology
  - computers
  - industrial automation
  - microelectronics
  - software
  - telecommunications
  - new materials
  - chemicals
  - aircraft
  - automotive

heavy electrical equipment.
- modes of cooperation, such as:[9]
  joint ventures
  joint R&D
  technology exchange agreements
  minority and cross-holding
  particular customer — supplier relations
  one-directional technology flows; and
- some comment, available information about progress.

Depending on the exact form of co-operation we collect information on:

- the operational context (national, combination of markets, worldwide);
- the name of the agreement, e.g. for joint ventures;
- equity-sharing;
- the direction of technology or capital flows;
- the degree of participation in a firm in case of minority holdings;
- some information about the motives underlying the alliance;
- the character of co-operation, such as: basic research, applied research, technology development, product development, production and/or marketing
- who will benefit most.

For each company we collect some general information such as:

- business activities;
- number of employees;
- turnover;
- profits;
- R&D expenditures.

Furthermore, we will include information about market structures, leading companies and number of US patents.

At the level of countries we can distinguish two groups of data:

- data that are directly provided by the information system itself (numbers of national and international agreements, public and private sector R&D agreements, number of companies); and
- indicators available at the national level from other sources.

A distinction is made in the following eight (supra-)national entities:

- USA
- Japan
- EC
- the remainder of Western European countries (Sweden, Switzerland, Finland, Norway, Austria)
- Eastern European countries (we registered some alliances having partners from the USSR, CSSR, Yugoslavia, GDR, Hungary and Bulgaria)
- other industrialised countries (Australia, Canada, New Zealand, South Africa, Israel)
- newly industrialised countries, so called NICs (Taiwan, South Korea, Singapore, Hong Kong, Brazil) and

developing countries (Argentina, People's Republic of China, India, Indonesia).

## Appendix II Classification of motives underlying technology co-operation agreements and their strategic impacts

In the following we will present a list of catchwords applied to analyse the major motives for co-operation in our present research. This inventory is based on the motives we found in the literature and in our previous research. These motives will be related to joint research and development pacts, joint ventures and technology-exchange agreements first, followed by a special list of additional motives for joint ventures and an inventory of motives for direct investments such as minority holdings. The list of motives is complemented by a number of forms of co-operation which by their nature have only one obvious purpose.

In addition to this we will relate each of these motives to their strategic impact. Throughout our research we will make a distinction between co-operative agreements which are aimed at the strategic, long-term perspective of the companies involved which we will indicate below with an '**s**', and those agreements which we think are more associated with the control of either transaction costs or operating costs of companies, which we will mark with a '**c**'. In case both general motives appear possible, either because it is not feasible to differentiate between the cost or the strategic argument or because partners can be expected to have alternating motives as a consequence of the character of the agreement; we have marked such agreements with a '**b**'.

*I. Motives, underlying joint research and development agreements, joint research pacts, joint ventures and research corporations:*

(a) costs and risks — high costs and risks of R&D in high-tech industries, **b** because both costs and strategic implications (risks) are relevant;
(b) the lack of sufficient financial resources, **b** although one might expect that costs are the main aspect of this motive, at least one of the partners will have a long-term strategic motivation in particular because most of these agreements are between a small company and a large financially supportive company;
(c) technological complementarity — increased complexity and inter-sectoral nature of new technologies, the technological competence of the partner, the necessity for some firms to monitor a spectrum of technologies (both offensive and defensive), **s**;
(d) reduction of innovation time-span — shortening the period between discovery (invention) and market introduction (innovation), getting patents as fast as possible, enlarge product supply, reduce period of development, **b** because both cost and strategic positioning are at stake;
(e) perform basic research, **s**;
(f) influencing market structure — setting up a combination against a third party (both offensive and defensive); reduce the number of competitors, competitive positioning, **s**;
(g) monitoring technological opportunities — monitoring the evolution of technologies and opportunities, switching to new — at least to the firm — promising technologies, **s**;
(h) hidden agenda — capturing and absorbing tacit knowledge, extracting skills, among other things by engaging into a network of changing partnerships, **s**.

*II. For joint ventures we will examine some further potential motives in addition to the ones mentioned above:*

(a) economies of scale, rationalisation of production, **c**;

(b) expansion, new markets — offensive marketing agreements, international expansion, internationalisation of market, exploitation of new market opportunities, **s**;
(c) defensive restructuring — outsourcing of peripheral activities, subsidiaries merger, **c**;
(d) national circumstances — benefiting from partner's national interests, **s**;
(e) bidding consortia, **s**.

*III Motives for direct investments, minority holdings and cross holdings:*

(a) tighten customer–supplier partnerships — to assure supply of high-quality components, sometimes in combination with contract-research, control of upstream activities and relations, in which costs control is decisive, therefore **c**;
(b) profitability — (potential) profitability of company to be invested in, **s**;
(c) matching of core activities — sharing particular core businesses in lateral relationship, **s**;
(d) monitoring possible entry — investigate possible entry into new fields of technology or new products, **s**;
(e) marketing entry — entry into marketing channel of the target company, as both stratetic and cost options can play a role: **b**;
(f) control over partner — to prevent a hostile take-over by a third party, to block would-be participators, create interlocking shareholder structure, **s**;
(g) technological competence of partner, **s**.

*IV Forms of co-operation which can be directly related to one dominant motive:*

co-makership contract — CMC, **c**;
co-production contract — CPC, **c**;
customer–supplier partnership — CSP, **c**;
licensing, **c** because it generates income for the licensor and reduces costs for the licensee;
combination of CSP and licensing, **b** because in such a combination the technology transfer involved is usually of a more strategic importance to at least one partner and also closer to best practice technology;
cross-licensing, **b**, see above;
research contract — RDC, **b** because it reduces the cost of research for one contractor but it also provides strategic information on technologies and capabilities of the other contractor;
second-sourcing agreement — SSA, **b**, both strategic options and costs of development lead companies to engage in such agreements;
mutual second-sourcing agreement — MSSA, **b**, see above;
technology sharing agreement — TS, as this agreement covers the sharing of existing technology cost economising is central, hence **c**;
combinations of TS with licensing, **c** see above.

In our assessment of the strategic implications of co-operative agreements between companies each agreement will be valued according to the list presented above. For agreements given under III one dominant motive can be accounted for. For other concrete agreements there are frequently several motives to be found of which some will be more strategic — and others more cost-related. In those cases such agreements will be valued as strategic, cost economising or mixed according to a 'common sense' interpretation of the occurrence and distribution of several motives per agreement.

Following such a procedure enables us to 'measure' the strategic implications of every agreement in our database.

**Appendix III List of company codes**

| *Code* | *Name* | *Country* |
|---|---|---|
| 3M | Minnesota Mining & Mfg. Co. | USA |
| ABB | ABB Asea Brown Boveri A.G. | SWI |
| ABBOTT | Abbot Laboratories Inc. | USA |
| AGS | Advanced Genetic Sciences | USA |
| AJINOM | Ajinomoto | JPN |
| AKZO | Akzo N.V. | NET |
| ALLIED | Allied Corp. | USA |
| AL-SIG | Allied-Signal Inc. | USA |
| AMAX | American Metal Climax (AMAX) | USA |
| AMCYAN | American Cyanamid Co. | USA |
| AMD | Advanced Micro Devices Inc. | USA |
| AMGEN | AMGen Inc. | USA |
| AMP | AMP Inc. | USA |
| ANNA-W | Anna Werke | FRG |
| ARCO | Arco (Atlantic Richfield Co.) | USA |
| ASAHI | Asahi Glass Co. | JPN |
| ASEA | Asea A.B. | SWE |
| AT&T | Am. Telephone & Telegraph Co. (AT&T) | USA |
| BASF | Basf A.G. | FRG |
| BAXTER | Baxter-Travenol Labs. Inc. | USA |
| BAYER | Bayer A.G. | FRG |
| BIOGEN | Biogen Inc. | USA |
| BIOLOG | Bio Logicals | CAN |
| BOLIDN | Boliden A.B. | SWE |
| BOSCH | Bosch | FRG |
| BP | British Petroleum | UK |
| BRIS-M | Bristol-Myers Co. | USA |
| BULL | Bull Groupe S.A. | FRA |
| BURROU | Burroughs | USA |
| CALGEN | Calgene Inc. | USA |
| CARDO | Cardo A.B. | SWE |
| CDC | Control Data Corp. (CDC) | USA |
| CELANE | Celanese Cor. | USA |
| CELTEC | Celltech Plc. | UK |
| CETUS | Cetus Corp. | USA |
| CGE | Cie. Générale d'électricité (CGE) | FRA |
| CHIRON | Chiron Corp. | USA |
| CHRYSL | Chrysler Motor Corp. | USA |
| CIBA-G | Ciba-Geigy A.G. | SWI |
| COLLAR | Collaborative Research Inc. | USA |
| CORNIN | Corning Glass Works | USA |
| DAIICH | Dai-ichi | JPN |

| | | |
|---|---|---|
| DAIMLR | Daimler-Benz A.G. | FRG |
| DEC | Digital Equipment Corp. (DEC) | USA |
| DEGUSS | Degussa A.G. | FRG |
| DNA-PT | DNA Plant Technology Corp. | USA |
| DOW | Dow Chemical | USA |
| DSM | DSM N.V. | NET |
| DUPONT | Du Pont de Nemours | USA |
| ELF | ELF Aquitaine S.A. | FRA |
| ELILIL | Eli Lilly & Co. | USA |
| ELKEM | Elkem A/S | NOR |
| ENIMNT | Enimont SpA. | ITA |
| ERICSS | Ericsson A.B. | SWE |
| FERRAN | Ferranti Plc. | UK |
| FERRUZ | Ferruzzi SpA. | ITA |
| FORD | Ford Motor Co. | USA |
| FR-FL | Friedrich Flick Industrie KGaA. | FRG |
| FUJITS | Fujitsu Ltd. | JPN |
| FUJI-E | Fuji Electric | JPN |
| GE | General Electric Co. (GE) | USA |
| GEC | GEC Plc. | UK |
| GENENT | Genentech Inc. | USA |
| GENEX | Genex Corp. | USA |
| GENZYM | Genzyme | USA |
| GEN-IN | Genetics Institute Inc. | USA |
| GIST-B | Gist-Brocades N.V. | NET |
| GM | General Motors | USA |
| GRCROS | Green Cross Corp. | JPN |
| GTE | General Telephone & Electric (GTE) | USA |
| HARRIS | Harris | USA |
| HERCUL | Hercules Corp. | USA |
| HITACH | Hitachi Ltd. | JPN |
| HLROCH | Hoffmann-La Roche & Co. A.G. | SWI |
| HOECHS | Hoechst A.G. | FRG |
| HOGANS | Hogans A.B. | SWE |
| HONEYW | Honeywell Corp. | USA |
| HYBRIT | Hybritech Inc. (Hybrid Technology) | USA |
| H-P | Hewlett-Packard Co. | USA |
| IBM | Int. Business Machines Corp. (IBM) | USA |
| ICI | Imperial Chemical Industries Plc. | UK |
| INCO | Int. Nickel Co. (Inco) | CAN |
| INTEL | Intel Corp. | USA |
| IRI | IRI | ITA |
| ISHIKA | Ishika-Wajima Harima Co. Ltd. | JPN |
| ITT | Int. Tel. & Telegraph Corp. (ITT) | USA |
| J&J | Johnson & Johnson | USA |
| KAWASA | Kawasaki | JPN |
| KKK | KKK | FRG |
| KODAK | Eastman Kodak Co. | USA |
| KOPPER | Koppers Co. | USA |
| KYOCER | Kyocera | JPN |
| KYOWA | Kyowa Hakko Kogyo | JPN |

| | | |
|---|---|---|
| LITTON | Litton Industries Inc. | USA |
| LOCKHD | Lockheed | USA |
| LUBRIZ | Lubrizol Corp. | USA |
| MAHLE | Mahle | FRG |
| MARTIN | Martin-Marietta Corp. | USA |
| MATSUS | Matsushita Elect. Industrial Co. Ltd. | JPN |
| MEIJI | Meiji Seika Kaisha Ltd. | JPN |
| MERCK | Merck & Co. | USA |
| MICSFT | Microsoft Corp. | USA |
| MITSUB | Mitsubishi Corp. | JPN |
| MITSUI | Mitsui & Co. | JPN |
| MOLGEN | Molecular Genetics | USA |
| MONSAN | Monsanto Co. | USA |
| MOTORO | Motorola Inc. | USA |
| NATSEM | National Semiconductor Corp. | USA |
| NCR | National Cash Register Corp. (NCR) | USA |
| NEC | Nippon Electric Corp. (NEC) | JPN |
| NISSAN | Nissan Motor Co. Ltd. | JPN |
| NIXDRF | Nixdorf | FRG |
| NOKIA | Nokia Oy. | FIN |
| NOVO | Novo Industri A.S. | DEN |
| NOVO-N | Novo-Nordisk A.S. | DEN |
| NSTEEL | Nippon Steel Corp. | JPN |
| NTT | Nippon Telegraph & Telephone (NTT) | JPN |
| NYBY | Nyby Uddeholm Powder A.B. | SWE |
| OKI | Oki Electric Industry Co. | JPN |
| OLIVET | Olivetti SpA. | ITA |
| PHILPS | Philips Gloeilampen-fabrieken, N.V. | NET |
| PLESSE | Plessey Co. | UK |
| PROCOR | Procordia Nova A.B. | SWE |
| RCA | RCA (Radio Co. of America) | USA |
| RHON-P | Rhone-Poulenc S.A. | FRA |
| ROSENT | Rosenthal A.G. | FRG |
| SANDOZ | Sandoz A.G. | SWI |
| SANDVK | Sandvik A.B. | SWE |
| SCAN | Scandust | SWE |
| SCHLUM | Schlumberger Corp./N.V. | USA |
| SHELL | Royal Dutch/Shell Group Plc./N.V. | NET |
| SIEMNS | Siemens A.G. | FRG |
| SKF | SKF | SWE |
| SMKB | SmithKline Beckman Corp. | USA |
| SPERRY | Sperry Corp. | USA |
| STC | Standard Telephone Co. Plc. | UK |
| STOIL | Standard Oil (Indiana) | USA |
| SUMITO | Sumitomo Group | JPN |
| SUN-M | Sun Microsystems | USA |
| SYNTEX | Syntex Corp. | USA |
| S-GOBN | Saint-Gobain S.A. | FRA |
| TANDEM | Tandem Corp. | USA |
| THOMSN | Thomson S.A. | FRA |
| TI | Texas Instruments Inc. | USA |

| | | |
|---|---|---|
| TORAY | Toray Industries Inc. | JPN |
| TOSHIB | Toshiba Corp. | JPN |
| TRW | Thompson Ramo Woolridge Inc. | USA |
| UNISYS | Unisys | USA |
| UN-TECH | United Technologies Corp. (UTV) | USA |
| VOLVO | Volvo A.B. | SWE |
| VW | Volkswagen A.G. | FRG |
| WESTIN | Westinghouse | USA |

## Notes

1. This paper is one of a series of papers in a research project on 'Inter-company Co-operation and Technological Development' at MERIT. This research focuses on the empirical analysis of changes in industry structures and global trends in different modes of inter-firm agreements in a large number of fields of technology. It also addresses theoretical questions in this field of research as well as methodological issues concerning applied network and multivariate analysis of strategies and industry structures. Research for this paper is partly financed with grants from the Ministry of Economic Affairs and the Advisory Group on Materials in the Netherlands. We thank David Mowery and Akira Goto for their comments on a first draft.
2. We define biotechnology as all applications of that particular field of technology in agriculture, pharmaceuticals, ecology, nutrition, chemicals and basic research. Information technologies are confined to computers, industrial automation, microelectronics, software and telecommunications. New materials are defined as new and improved electronics materials, technical ceramics, fibre-strengthened composites, technical plastics, powder metallurgy and special metals and alloys.
3. In our CATI databank there is still a very strong bias as far as the number of intra-Japanese agreements is concerned. We are quite confident about the quality of our material on Japanese collaboration with others, but for many reasons, not least the language barrier, we still lack sufficient information about co-operative agreements between Japanese companies.
4. We have chosen to give this distribution for the sums of motives being mentioned, adding single mentioned motives to motives which were referred to as one of a number of motives for particular agreements. Choosing either the sequence of single mentioned motives or the sequence of motives mentioned in a combination does not affect the overall ranking of motives. This led us to take the total number of times a motive is mentioned as the main indicator for the relevance of a particular motive. Consequently, the totals in Tables 3, 4, 5 exceed the number of agreements.
5. See Appendix II for additional informtion in the procedure followed to associate each agreement with one of these alternative strategies.
6. In these two fields between 55 and 60 per cent of the agreements are strategically motivated, pure cost economising agreements have little relevance.
7. These similarity matrices are not reproduced in this paper.
8. There are a number of sub-fields for most of these technologies.
9. Each mode of co-operation has a number of particular categories.

## References

Hagedoorn, J. (1989a), *The Dynamic Analysis of Innovation and Diffusion*, London: Pinter Publishers.

Hagedoorn, J. (1989b), 'Partnering and Reorganization of Research and Production: Global Strategies in Manufacturing', Umea, working paper CERUM.

Hagedoorn, J. (1989c), 'Theory and Analyses of Partnerships in Production and Innovation' working paper, MERIT.

Hagedoorn, J. (1990), Organisational Modes of Inter-firm Co-operation and Technology Transfer, in *Technovation* 10, pp. 17–30.

Hagedoorn, J. and J. Schakenraad (1989), 'Strategic partnering and technological co-operation' in Dankbaar, B., Groenewegen, J. and H. Schenk (eds), *Perspectives in Industrial Economics*, Dordrecht: Kluwer.

Haklisch, C.S. (1986), 'Technical Alliances in the Semiconductor Industry', mimeo, NYU.

Hergert, M. and D. Morris (1986), 'Trends in International Collaborative Agreements', INSEAD paper.

Hladik, K.J. (1985), *International Joint Ventures*, Lexington: Lexington Books.

Kay, N. (1979), *The Innovating Firm*, London: Macmillan.

OECD (1986), *Technical Co-operation Agreements Between Firms: Some Initial Data and Analysis*, Paris: OECD.

Ohmae, K. (1985), *Triad Power*, New York: Free Press.

Riordan, M.H. and O.E. Williamson (1985), 'Asset specificity and economic organization', *International Journal of Industrial Organization* 3, pp. 365–378.

Tulder, R. van and G. Junne (1988), *European Multinationals in Core Technolgies*, Chichester: John Wiley.

Williamson, O.E. (1985), *The Economic Institutions of Capitalism*, New York: The Free Press.

# 2. The formation of firm strategy as self-organisation[1]

*Georges Romme*

Everything about organisational strategy conveys dynamics. Yet, few organisation theorists have devoted attention to strategic change (e.g. Burgelman, 1983; Mintzberg and Waters, 1985) and even less attention has been given to the source of change or how it comes about. This chapter offers a framework for longitudinal analysis of the formation[2] of organisational strategy. The self-organisation metaphor is put forward as a useful image in describing and explaining strategy formation. The methodology involves generating theory from the perspective of the self-organisation metaphor. That is, broad categories of observations and existing theories will be recast in a self-organisation framework.

As regards structure, I start with a short comment on methodology before making some brief comments on the state of the art of strategy formation research. Then, the focus switches to several elementary assumptions of the self-organisation metaphor, and a self-organisational theory of strategy formation is outlined. The resulting set of propositions is grounded in a number of observations from strategy research.

## Methodology

The methodology of discovering grounded theory was developed by Glaser and Strauss (1967). The grounded theory methodology involves generating theory through existing data and field studies. That is, broad categories of data and existing theories are analysed, with a primary emphasis on theory generation instead of theory testing. In attempting to 'discover' theory, according to Glaser and Strauss (1967), one has to generate conceptual categories or their properties from observations, and then use the observations from which a category emerges to illustrate the theory.

However, the development of theory is a two-sided process, involving 'grounding' as well as 'imaginisation' (Morgan, 1986). Conceptual categories arise from observation as well as from a way of seeing or thinking. The latter element in the development of theory implies that our theories are based on metaphors that lead us to see and understand phenomena in distinctive yet partial ways. Grounding in empirical observation and imaginisation by means of metaphors are two quite autonomous processes. That is, one cannot infer metaphors from observations, or vice versa.

The importance of metaphors for understanding 'organisation' in its broadest sense has been explored by Morgan (1986). According to Morgan, a metaphor always produces a kind of one-sided insight. In highlighting certain interpretations, it tends to force others into a background role. Extensive attention should be given to relevant metaphors as much as possible and as early as possible in the process of theory development. That way, insight is gained into the potential and limitation of theory. The methodological premise in this paper therefore holds that in relatively unstructured research fields priority should be given, first, to metaphor(s) worth elaborating; and, second, to grounding concepts and theory. Consequently, this paper will look initially at the metaphor(s) conducive to theory development in the field of strategy formation. Then five primary conceptual categories are put forward as the skeleton of a self-organisation framework. Finally, the self-organisation framework will act as a focusing device in generating a number of propositions on the basis of typical observations from research into the formation of strategy.

## Strategy formation as self-organisation

The dominant transformation in organisation theory from closed to open system views and theories (Scott, 1987) is also reflected in strategy formation research. The contingency metaphor has become dominant in most conceptual and empirical research (e.g. Mintzberg, 1973; Harrison, Torres and Kukalis; Hickson *et al.*, 1988). The contingency literature focuses on the conditions favouring specific kinds of strategy modes. It emphasises situational differences and leaves underlying organisational and managerial processes to be inferred (Chaffee, 1985; Romme, 1990a).

Recently, several authors have returned to the more traditional closed system view (Mintzberg, 1978; Mintzberg and Waters, 1985; Burgelman, 1983; Romme, 1990a). Romme (1990b) suggests that the common ground in these studies can be understood in terms of the self-organisation metaphor. The following assumptions serve as an outline of strategy formation as a self-organising system (Romme, 1990b).

First of all, the self-organisation metaphor assumes that *strategy forming systems are closed and open at the same time*. Strategy (forming systems) maintain their existence by opening up in particular ways to the outside world. The way they open up, however, relies on a closed system of actions *and* interpretations. In short, strategy systems open up to environmental imperatives in ways that are determined in the strategy system itself.

*Strategy systems are complex.* The behaviour of (elements of) a strategy system cannot be induced from its inputs or its internal conditions, but is the product of the interaction between both.

*Strategy systems are self-referential.* That is, such systems are 'operationally closed'. Any behaviour of (actors within) the system feeds back on itself and forms the point of departure for subsequent behaviour.

Finally, *strategy systems are autonomous.* A system is autonomous if the relations and interactions which define the system as a whole merely involve the system itself and no other system. This is an autonomy relative to certain criteria: the system should not be administered, produced or developed by external entities. Thus, a system can be autonomous and at the same time be dependent on external influences (e.g. resources, markets or technologies).

The self-organisation idea suggests that the question whether organisations are closed or open systems is not interesting. Instead, organisations tend to maintain their existence by opening up in particular ways to the 'outside world'. The sharp dichotomy between organisations and environments is misleading. Environments are as much a part of organisations as are organisational structures, administrative activities or production facilities. Organisations are thus viewed as circular patterns of interaction whereby change in one element of the system is coupled with changes elsewhere, setting up continuous patterns of interaction (Morgan, 1986). This kind of circular reasoning may indeed seem very strange (Morgan, 1986). However, Maturana and Varela (1980) argue that this is because we insist on understanding these systems from our point of view as observers, rather than attempting to understand their inner logic. If we put ourselves 'inside' such systems we may come to realise that we are within a closed system of interaction and that the environment is part of the organisation because it is part of its domain of essential interaction.

## Categories of strategy formation

Strategy formation contains many self-organising phenomena. Several studies of strategic change have observed a broad variety of processes in which old and new ways of understanding and behaviour interact (e.g. Kanter, 1983; Bartunek, 1984). It is therefore important to consider which theoretical categories are primary and which are subsidiary (Morgan, 1986). I assume that interpretation and action are the two major elements of any strategic change process. The interpretation side involves the problem of sensing and (re)cognising and subsequently inducing and driving action, in other words, providing a direction. The action side involves actual behaviour of actors participating in, contributing to or influencing the strategy process.

In looking at strategy formation I adopt the view of organisations as social groups, whose members are attempting to adapt and survive in their particular circumstances.[3] Within each organisation, dominant groups (Cyert and March, 1963; Penrose, 1959) have to obtain some cohesion in order to impart a particular character and direction to organisational activity and develop particular ways of working together. Dominant groups give organisations their cohesive character and generate the, often tacit, distinctive repertoire to cope with difficulties. Thus, strategy formation is viewed as a group process. In real-world cases the strategy making context may range from extremely tight groups to very loose collections of individuals or other groups.

In the remainder of this section I will describe five primary categories which specify action and interpretation in strategy formation by dominant groups. These categories are:

(a) difficulties of the dominant group;
(b) perception by the dominant group;

(c) repertoire of the dominant group;
(d) actions induced by the dominant group; and
(e) actions autonomous to the dominant group.

Environments are reflections and part of strategy formation itself. Thus, strategy formation cannot enter into interactions that are not specified in the pattern of relations defined by itself. Strategy makers interact with their environments in a way that facilitates their own self-organisation as strategy makers. The strategic uncertainties (e.g. problems, opportunities) the environment poses to the dominant group are part of the strategy formation process. These strategic uncertainties will be referred to as *difficulties* or the difficulty set.

Another primary category is *perception* by the dominant group. Two activities are pertinent to the collective perception of the dominant group: sensory and cognitive activity (Romme, 1990a). Sensory activity or sensing involves exposure to difficulties as raw data, e.g. by search, exploration or informal contacts. Sensing relies on some looseness of group perception as it especially involves the individual exposures to raw data. Cognitive activity or sense-making pertains to the interpretation and understanding of available information. Although sense-making in many circumstances may be predominantly an individual activity, in the context of group perception I will focus on the sharedness of individual cognitive frameworks. Thus, I assume that sense-making, in contrast to sensing, is conditioned by some tightness of group perception. Every dominant group thus has to obtain some cohesion in individual perceptions in order to constitute some consensus in cognition as the basis for producing repertoires of action (Nystrom, Hedberg and Starbuck, 1976). The perceptual cohesiveness of a dominant group may be defined in terms of its resistance to disruptive forces (cf. Gross and Martin, 1952). The essence of the group learning process implied in this view is that individual sensing must be complemented with cognitive activity which encourages some level of shared problem understanding, if change is not to be imperilled at birth (Pettigrew, 1985).

Note that group perception and group formation, at least in the case of organisational strategy, are intimately related categories. An individual entering a strategy formation process at first sees himself in an environment which is relatively indifferent to his individuality but which 'implicates him in a web of circumstances' (Simmel, 1955). Group membership develops when the coexistence with others tends to associate homogeneous members from (multiple) heterogeneous groups (Simmel, 1955). As perception by a dominant group relies on both individuality and collective sense-making, we can view group perception and group formation as two analogous, related processes, which are influenced by the same set of forces.

Repertoires are perception-action translations giving organisations a sense of direction. Their role is to drive decisions and actions in the absence of more deliberate modes of addressing difficulties (Romme, 1990a). *Repertoire*, here, refers to a collection of rules prevailing in a dominant group at a certain moment in time (Romme, 1990a). Repertoires comprise rules related to strategy content, e.g. with regard to positioning in market, technology and industry, as well as strategy process, e.g. with regard to decision making conditions. Typically, a repertoire leaves room for a wide range of more specific intentions but provides enough structure to guide actions. Repertoires do not necessarily involve clear-cut, explicit or conscious guidelines for organisational action. For instance, one may employ more tacit, global and unconscious rules, possibly originating from current industry practice (e.g. 'in our business, co-operation between competitors is not done').

The proof of the pudding for any strategy lies in action. It is useful to make a distinction between *actions* which are '*induced*' by the repertoire and '*autonomous*' ones. A similar distinction is made by Burgelman (1983), who refers to autonomous action as introducing new categories for the definition of opportunities. Certain actions will appear as autonomous relative to others being deliberately induced by the dominant group as a whole. Repertoires and induced action co-produce autonomous action (I will be more specific in a later stage). A position of extreme subjectivism is not taken here. It is merely assumed that a dominant group engages objective reality under the conditions of its own repertoire and induced actions. Typical examples of induced action are investments in facilities, new product development for existing business and planned take-overs. Examples of autonomous action include product championing efforts, internal political moves changing the composition of the dominant group, interventions by shareholders, major initiatives by competitors or customers, and grass-roots groups of activists mobilising to protest against policies of the dominant group.

The relative weight of induced and autonomous action can be viewed in terms of the controllability of action from the perspective of the dominant group. Of course, autonomous actions can be undertaken by members of the dominant group (e.g. Chief Executive Officers), actors on the periphery of the group (e.g. marketing managers) as well as actors outside the group (e.g. competitors). Autonomous action opens up the difficulty of the problems to be addressed. Difficulties, in turn, evoke sensing and sense-making in order to close towards a repertoire for action. This part of the strategy formation cycle is central to the enactment argument (Weick, 1979), especially when difficulties arise from autonomous actions by members of the dominant group.

## Propositions on forces in strategy formation

A number of propositions specifying strategy formation as a self-organising system will now be advanced. They are generated by viewing a wide set of observations in strategy formation from the perspective of the self-organisation metaphor. For the purpose of overview, the ten relations in Figure 1 can be characterised as follows:

(1) difficulties evoke sensing but impede sense-making;
(2) perception generates repertoire;
(3) repertoire induces action;
(4) induced action generates autonomous action;

(5) autonomous action produces difficulties;
(6) difficulties activate repertoire;
(7) repertoire blinds autonomous action;
(8) autonomous action affects cohesion of perception;
(9) looseness of perception distracts induced action; and
(10) induced action diverts attention from difficulties.

The relations in Figure 1 can be conceived as tendencies or forces driving strategy formation. The first five relations pertain to rather complementary forces around the cycle, whereas the last five focus on more antagonistic forces across the cycle. I will now turn to more specific assessments of the self-organisation forces listed above. The verb 'open (up)' indicates that a category broadens in range to include certain interpretations or actions. Vice versa, 'close' indicates the narrowing of a category by eliminating certain interpretations or actions.

*Difficulties evoke sensing but impede sense-making.* Sensing is facilitated by loose perception of the dominant group, whereas collective sense-making is stimulated by especially high cohesiveness of perception. Case-studies (e.g. Starbuck, 1983; Kanter, 1983; Pettigrew, 1985; Bartunek, 1984) include numerous examples of managers, leaders or other parties contributing to strategy formation by sensing major difficulties. Overall, these cases suggest that sensing is predominantly an individual activity, countering group tendencies to

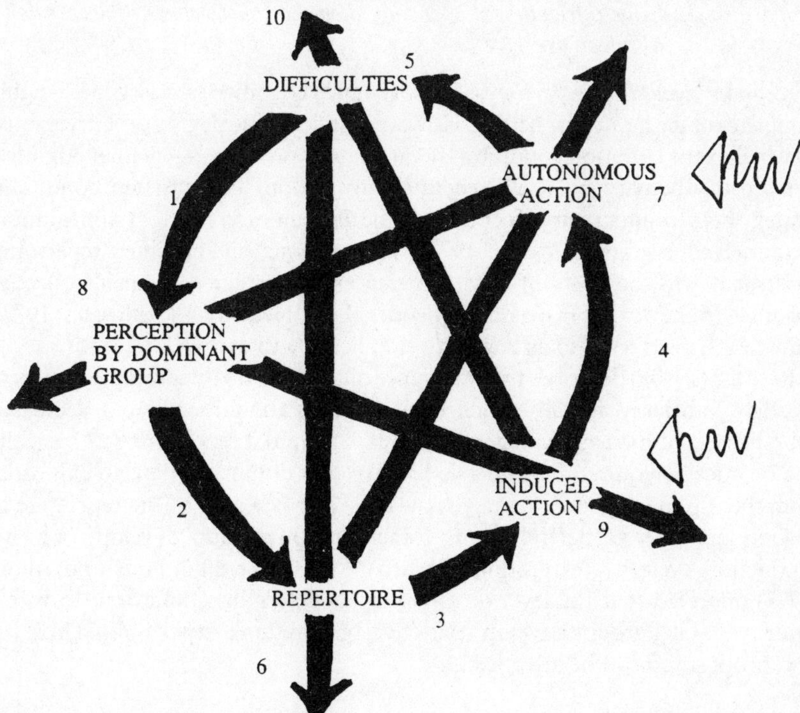


**Figure 1** Self-organizing forces in strategy formation

produce tight, shared understandings of the world outside. For instance, the pervasive role of task forces at Honeywell or management committees at J.C. Penney — with a variety of fields and functions represented and diverse in race, gender and culture — clearly contribued to sensing activity of the dominant group as a whole (Kanter, 1983). This means that a dominant group whose composition is diverse and flexible is more likely to pick up on external cues than a dominant group with a smaller, more homogeneous set of actors or with a single function, whether finance, marketing or any other, having disproportionate power to define the appropriate focuses of attention (Kanter, 1983).

Observations of shifts in top management positions provide a more specific perspective. Shifts in CEO and chair of the board positions have been studied by Harrison, Torres and Kukalis (1988). They show that generally a large difficulty set (e.g. poor performance of the company) can reduce CEO and chair influence and lead to a (further) separation of the positions, including the appointment of a new CEO. Vice versa, a limited difficulty set (e.g. strong performance) enhances CEO power, possibly leading to a consolidation of the two top positions under the current CEO. These findings suggest that major difficulties such as a crisis of sufficient magnitude can shift the locus of power temporarily to the board, or more generally, reduce the overall cohesion of the dominant group, reflected in a substitution between collective sense making and individual sensing. Thus, it appears plausible to put forward the first proposition.

*Proposition 1: If the difficulty set opens up, then the cohesiveness of perception will decrease*

*Perception generates repertoire.* Sensory and cognitive activity are highly complementary inputs for action. However, under some degree of 'uncertainty' — which is the common condition for all strategy processes — neither cognitive nor sensory activity by itself generates any action. In uncertain conditions strategy makers alter their perceptions of the difficulties so that they appear more certain (Hedberg and Jönsson, 1977). They will act on simplified repertoires constructed on the basis of past experiences accumulated in their collective cognitive framework (Grinyer and Spender, 1979; Hedberg and Jönsson, 1977), a process also referred to as cognitive simplication (Schwenk, 1984).

Romme (1990a) describes the emergence of the forward integration strategy of Hendrix Fabrieken in the Netherlands as being conditioned by a decreased cohesiveness of the top management team. The initial repertoire was basically one of 'stick to your core business'. The exit of a dominant, patriarchal leader from the top management team gave way to a broadening of the repertoire to include such rules as 'defend against foreclosure threat' and 'develop your own production cycle by integrating forward'. More generally, Janis and Mann (1977) observed that the level of cohesiveness determines the extent to which strategy making groups develop restrictive, 'group think' repertoires. Thus, the next proposition can be advanced.

*Proposition 2: If the cohesiveness of perception increases, then the repertoire will close.*

*Repertoire induces action.* A repertoire generates action by enabling the agents to action, e.g. by giving rules for the allocation of resources or delimiting certain domains of managerial behaviour (such as 'do not become involved with union representatives'). Starbuck and Hedberg (1977) describe the case of Facit AB, a company based in Sweden that grew large while making and selling business machines and office furnishings. Although Facit made many products, in the course of time top management's repertoire steadily closed down around the rule that the key product line should be mechanical calculators. Thus, Facit's actions focused on the production and sale of mechanical calculators by improving the quality and lowering the costs. For similar illustrations, see Kanter (1983) and Romme (1990a). Anticipating the activation phenomenon (to be discussed in the context of proposition 6), the third proposition holds that a broadening of the repertoire will induce less restricted actions.

*Proposition 3: If the (activated part of the) repertoire opens up, then the induced actions will be less restricted.*

*Induced action generates autonomous action.* Induced action may produce satisfactory or even excellent results, but only within the domain of the repertoire (Starbuck, 1983). Imagine the development from a small, entrepreneurial firm to a larger, multidivisional company. In the entrepreneurial stage, induced actions are typically rather strong and restrictive. Autonomous action by, for instance, competitors or customers generally reflects a high change level (e.g. major customers demanding another market approach). In the context of large, complex companies, induced action on the level of the head office is generally less restrictive, due to the impossibility of getting acquainted with details in the broad spectrum of activities. Burgelman (1983) observes that in large companies induced action intervenes only to a limited extent in the way autonomous action influences the strategy formation process. Thus, more restricted, induced actions will be paralleled by a wide range of autonomous actions, and vice versa.

Consider again the case of Facit AB. Facit's actions focused on mechanical calculators instead of the emerging electronic calculators, for instance, by relegating its electronic engineers to a small, jointly owned subsidiary. Top, middle and lower managers agreed about how a mechanical-calculator factory should operate, what customers wanted, etc. Also, no resources were wasted gathering 'irrelevant' information. The resulting autonomous actions were quite dramatic. For instance, one loyal customer cancelled a large order after Facit had failed repeatedly to produce machines of adequate quality; also, overall demand for mechanical calculators dropped precipitously and the board of directors sold Facit to a larger firm (Starbuck and Hedberg, 1977). Note that the restrictiveness of induced action relative to autonomous action is in fact the controllability of organisational action as a whole. Proposition 4 suggests that the controllability of organisational action increases if induced actions are less restricted.

*Proposition 4: If the induced actions are less restricted, then autonomous action will be more restricted.*

*Autonomous action produces difficulties.* The self-organisation framework implies that difficulties are generated by autonomous action inside, on the periphery of, or outside the dominant group. For instance, inside the group, vacancy difficulties may arise because of the exit of a top manager; in the periphery of the group, sales difficulties may be produced by loyal customers cancelling large orders (Starbuck and Hedberg, 1977) or input difficulties may arise because of the resistance of a major supplier to redesign basic inputs (Romme, 1990a); over-investment difficulties may be generated by actions at some distance from its top management, e.g. competitors contributing to industrial over-capacity, or by a varied set of actions constituting a severe recession and currency problems (Pettigrew, 1985). In short, the fifth proposition may be stated as follows:

*Proposition 5: If autonomous action is less restricted, then the difficulty set opens up.*

The first five propositions show that strategy formation in its more 'natural' mode develops according to so-called deviation counteracting loops (Masuch, 1985). For instance, cohesive groups tending towards group think behaviour (Janis and Mann, 1977) will in the long run produce autonomous actions and difficulties which will open up the closure of the group.

*Difficulties activate repertoire.* Repertoires contain both dormant and active rules (Clark and Starkey, 1988). Dormant rules can be activated and active rules can develop into dormant ones if no longer activated. An obvious condition for activation to occur seems to be a relatively stable composition of the dominant group. That is, major shifts in the composition of the dominant group may impede the activation of dormant repertoires employed by previous occupants of the positions in the group.

A clear example was observed in the way a dormant repertoire of one major management team in the 1960s at Corah (a major garments supplier of Marks & Spencer) was activated and brought into action (Clark and Starkey, 1988). In the case of Corah's main managerial team, anticipation of fundamental shifts in consumer preferences led to a long dormant repertoire in designing garments being activated. By contrast, a relatively new, quasi-autonomous production unit of Corah, administered by a management team that did not include any long-serving manager with expertise in marketing, did not succeed in a similar attempt. The lack of a dormant repertoire which includes guidelines for coping with market change in the knitted goods industry appears to be one major explanation (Clark and Starkey, 1988). Facit's development illustrates the case of an apparently stable difficulty set continuously re-activating a well-established repertoire, i.e. of producing and marketing mechanical calculators. Again and again, Facit's top management responded to low financial performance by, for instance, laying off employees, major reorganisations, replacing individual managers and closing plants (Starbuck and Hedberg, 1977). A major response to difficulties may thus involve trying out rules already in the repertoire, which may interfere with efforts to develop more appropriate rules. In sum, I suggest the following proposition.

*Proposition 6: If the difficulty set opens up, (dormant) parts of the repertoire are activated.*

*Repertoire blinds autonomous action.* Repertoires are dual in nature. That is, in addition to enabling action (see proposition 3) they also constrain the agents of action to operate (Giddens, 1979). In this respect, it is noteworthy to observe that some of Peters and Waterman's 'excellent' companies have not been able to extend their excellence over longer periods of time (Soeters, 1986). Their well-defined, strong repertoires — so strong that one either buys into their norms or gets out — apparently contribute to the risk of becoming myopic. For example, Levi Strauss and Co. held on too long to its 'reality' of successful jeans. It reacted too late when the preferences of a majority of the customers shifted away from jeans, resulting in a quick downfall in the financial results (Soeters, 1986). The strong, ingrained and often egocentric repertoires of the leadership of such organisations may severely constrain autonomous action and thus they truncate and distort their understanding of the wider context in which they operate, and surrender their future to the way the context evolves (Morgan, 1986). Similarly, Janis and Mann (1977) observe that cohesive decision-making groups produce narrow repertoires, in extreme cases resulting in isolated and defensive group behaviour.

Case studies of organisations facing crises provide clear examples (Nystrom, Hedberg and Starbuck, 1976; Starbuck, Greve and Hedberg, 1978; Starbuck and Hedberg, 1977). Narrow repertoires focus perception on the (especially induced) actions that their creators believe important, so the repertoire blinds the dominant group to other actions that often turn out to be more important (Starbuck, 1983). We can understand this blinding phenomenon as autonomous action moving away from the dominant group. Such movement may take place in time as well as in space. In terms of time, autonomous action may shift from the present time to the future. In terms of space, autonomous action may change scenes from the core to the periphery of the dominant group and, ultimately, to other actors.

Again, the development of Facit AB presents dramatic examples (Starbuck and Hedberg, 1977). The highly focused repertoire of Facit's top management prevented its members from undertaking any explorations with regard to the rise of the electronic calculator. In a way, such autonomous action was delayed into the future. Also, autonomous action moved from lower-level managers and engineers, who were signalling the electronic revolution in the world at large but did not succeed in bringing such awareness upward, to a loyal customer cancelling a large order and, finally, a larger firm taking over Facit. In sum, the advent of electronic calculators took Facit's top management by surprise. This surprise, however, is clearly (partly) conditioned by the very narrow repertoire they developed.

*Proposition 7: If the repertoire opens, then autonomous actions move (in time and space) towards the dominant group.*

*Autonomous action affects cohesion of perception.* Where groups or organisations appear to 'act', there are often strong individuals persistently pushing (Kanter, 1983). Autonomous action supports the sensing capacity of dominant groups (especially if it involves actions by members of the group), but tends to harm collective sense-making. Think of the distorting effect crisis situations may have on management teams. In this respect, Fiol and Lyles (1985) observe that extremely high action change levels parallel low cognitive levels. That is, in extremely turbulent situations management is not capable of group learning.

Research into the role of the CEO and the chair of the board of directors may put things into perspective. Harrison, Torres and Kukalis (1988) infer from their data that the exit of a dominant executive holding both the CEO and the chair position generally reduces the concentration of power in the dominant group and can lead to a separation of the two top positions.

Another example of the relation between autonomy of action and collective perception is provided by Mintzberg *et al.* (1986). In Arcop, a Canadian architecture co-partnership with a record of 'excellence in craft', the looseness of the group of partners was clearly constituted by autonomous action by partners, competitors and clients, reflected, for example in the diverse initiatives of individual partners and their attempts to carve out their own domains. In the history of Arcop, the exit of a lot of persons (including partners) from the firm constituted a critical landmark. Such major changes in its size and composition may involve the complete disintegration of the group. Of course, such a disintegration does not necessarily bring along a disintegration of the organisation as a whole. Other forces contribute to the scene in which either the organisation breaks down (e.g. bankruptcy) or a new dominant group evolves and steps into the vacuum. In Arcop's case, the remaining part of the group took command (Mintzberg *et al.*, 1986). Other cases provide similar examples of the following proposition (Bartunek, 1984; Kanter, 1983; Mintzberg and McHugh, 1985; Starbuck and Hedberg, 1977; Wilson, 1982).

*Proposition 8: If autonomous actions move (in time and space) towards the dominant group, then the cohesiveness of perception decreases.*

*Looseness of perception distracts induced action.* Groups low in cohesion will have to deal more with facilitating collective action (if any) at the cost of action itself. Consider two cases: extremely low and extremely high perceptual cohesiveness. The organisational context known as the 'garbage can' arises in dominant groups with extremely low perceptual cohesion, reflected in a variety of inconsistent, ill-defined preferences and fluid participation (Cohen, March and Olsen, 1972). Actions in a 'garbage can' situation involve various kinds of problems and solutions being dumped by participants as they are generated (Cohen, March and Olsen, 1972). Several case-studies provide colourful illustrations of 'garbage can' actions in the form of power exertion or political manoeuvres serving local interests (Kanter, 1983; Pettigrew, 1985; Wilson, 1982).

The formation of what may be called ideological strategies (Mintzberg and Waters, 1985) provides examples of the implications of extremely high cohesion (e.g. Bartunek, 1984; Romme *et al.*, 1990). The actions observed in such cases are directly driven by a collective ideology-based perception rather than by a mediating repertoire.

Research on roles involved in the formation of innovation decisions (Schon, 1963; Maidique, 1980; Rothwell *et al.*, 1974) provides another perspective on the ninth relation. The importance of the various roles evolves as a function of the stages of business development, including the structure of the dominant group. In the small firm, the entrepreneur, for example, helps to define new products while still maintaining control of the overall organisation. When the firm grows larger and the dominant group becomes less cohesive, the entrepreneur gives up technical definition and sponsorship of most new products or processes. Sponsorship of projects that emerge from the technical department is passed on to, for instance, product championing middle managers, with the entrepreneur having final approval. On arrival at the stage of the loosely coupled, diversified company, new roles are again created to bridge the gap between the entrepreneur and the champion (Maidique, 1980). If we adopt the definition of political action as organisational behaviour that (attempts to) influence the distribution of advantages and disadvantages among local interests (cf. Robbins, 1984), the following proposition can be advanced.

*Proposition 9: If the cohesiveness of perception increases, then induced action includes less action of a political kind.*

*Induced action diverts attention from difficulties.* In many owner-controlled, entrepreneurial firms we observe the following phenomenon: after the start-up problem has more or less been solved, entrepreneurs continue to focus on the initial difficulty set (e.g. Mintzberg and Waters, 1982). The durability of such 'successfully reduced problems' can also be inferred from the history of Facit, whose management focused on the development, production and sales of mechanical calculators, although other major difficulties gained prominence. Facit's top management, for example, repeatedly announced that the firm was sound and an upturn was imminent. Until two days before Facit was sold such announcements were issued. Although the top management group probably believed them, these announcements voiced unrealistic hopes; to other personnel they may have been credible until the end (Starbuck and Hedberg, 1977). Other examples are revealed in studies of the US government's escalation of military activity in Vietnam in the early 1960s, involving dramatic consequences of 'over-realised' actions for the perceived difficulty set (Mintzberg, 1978; Mintzberg and Waters, 1985).

This phenomenon can be viewed as induced action reducing some difficulties and thereby diverting attention from actual or new difficulties. That is, organisational action as a whole seems to be under the control of the dominant group. The attention to past success may translate into records of excellence in terms of managerial repertoire and action merely if the other force opening the difficulty set (see proposition 5) is of minor strength or can be suppressed.

Particularly vulnerable are organisations that must commit large quantities of resources to specific missions in the context of, apparently, stable difficulty sets. In sum, 'solving' major difficulties by effective action interferes with the opening of the difficulty set.

*Proposition 10: If induced action is less restricted, then the difficulty set opens up.*

The addition of propositions concerning antagonisms in strategy formation gives way to the phenomenon of deviation amplifying loops (Masuch, 1985). That is, whereas the first five propositions constitute a deviation counteracting system, the last five relate to forces amplifying local imbalances, by aggravating or tightening existing imbalances. Amplification of local imbalances may occur as a result of time lags between the forces at work or as a result of rigidity in one of the categories. For instance, the diversion force (proposition 10) may contribute to the tendency of a rather cohesive group to produce a restrictive repertoire and action (propositions 2 and 3) by diverting attention from emerging difficulties. A closer look at strategy formation in Facit AB provides a number of examples of antagonistic forces amplifying local imbalances.

For the sake of overview, the development of Facit's strategy is divided into three periods, in each of which certain forces appear to dominate. In period one (see Figure 2a), a rather cohesive dominant group develops with Facit's top managers at its core. Having solved the entrepreneurial difficulties of setting up facilities for the production and marketing of mechanical calculators and related products, the group's repertoire closes down around one central rule: our business is in mechanical calculators. The actions which are induced reflect this restrictiveness. At the same time, autonomous action drifts away from Facit's management (from marketing managers to customers and to competitors), thus constituting the experience of a stagnating environment.

In period two (see Figure 2b), the initial imbalance between strong closing and

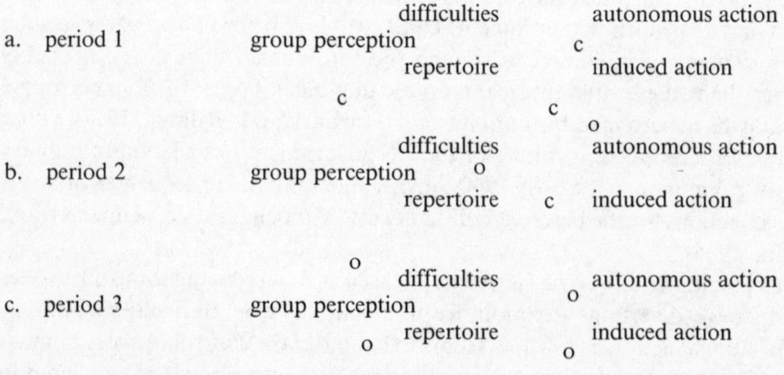

a. period 1
difficulties autonomous action
group perception c
repertoire induced action
c c
o
b. period 2
difficulties autonomous action
group perception o
repertoire c induced action
o
c. period 3
difficulties autonomous action
group perception o
repertoire induced action
o o

**Figure 2** A schematic interpretation of strategy formation at Facit AB.
(c refers to a closing force and o to an opening force)

weak opening forces is tightened and deepened. Autonomous actions are now taken predominantly by external agents with increasingly excessive effects (cancelling large orders, threat of substitute product). The potentially opening impact of these developments on the perceived difficulty set is reduced as a result of induced actions which are defending current prices, thereby diverting attention from new threats and opportunities. Moreover, top management relapses into re-activating (or opening) rules already in the repertoire, at the cost of any opening of the repertoire towards new, more appropriate actions.

In period three (see Figure 2c), forces develop which restore the amplified imbalance. Top managers start to seek new owners for Facit. Within a week after the new owner, Electrolux, has taken full control, Facit's chief executives are evicted, and responsibilities for Facit's operations are distributed among Electrolux's top management. Thus, the dominant group is infused with more realism and the need for strategic reorientations is perceived. The new dominant group delivers 'successful' action by financing new developments, reducing the need to act solely out of financial emergency, converting existing plants to other products, etc. (Starbuck and Hedberg, 1977).

## Conclusion

The development of theory is a continuous process. Therefore, each theory should be regarded as a momentary product. The theoretical framework developed in this paper provides an integrative perspective on the work of many strategy reserchers. None the less, as a whole it still lacks the support of a broad set of data. Systematic case-studies across a number of countries, industries and periods are scarce and will be needed in order to arrive at a well-codified set of propositions. Rigorous tests may prove to be very difficult. Large databases cannot provide the necessary material because of the essentially longitudinal character of the variety of forces. Moreover, the simultaneous influence of different forces on strategy formation poses additional problems. Further development and testing will therefore have to rely on in-depth case studies, using process methodologies (Mohr, 1982) which track the behaviour of dominant groups over longer periods of time.

The self-organisation metaphor may provide a new lens on the state of the art of strategy formation research. It is well suited for stripping away surface events and revealing underlying forces in strategy formation. Subsequent studies will have to deal more extensively with the conjunction of primary forces and contingent combinations of circumstances. They may show how forces orchestrate the conditions of strategy formation and how contingent combinations of surface phenomena channel and shape the actual realisation of forces.

I have focused on five categories giving rise to ten forces in strategy formation, discarding a variety of other phenomena. This rather restricted attention is necessary because primary categories may be masked by other, subsidiary categories which, for instance, may be more visible. Subsidiary tensions tend to arise within the undercurrents of self-organisation. For example, the various 'stakes' different participants have can be treated as subsidiary categories masking the cohesiveness of perception as a primary category. Under conditions of loose perception one may understand impediments in strategic decision making as being caused by high contention between interests or lack of effective communication. Of course, both researchers and the actors themselves are vulnerable to this 'masking' bias. In any case, the assertion of such bias is merely valid as long as the self-organisation metaphor guides our attention.

The major weakness and strength of the self-organisation metaphor concur in its one-sided view of the world in terms of forces always evoking other forces. There is no assumption of an environmental impetus for change to begin. Instead, it is assumed that autonomous actions and difficulties are engaged under the conditions of the incumbent group's actions and repertoire. Thus, some evolution is occurring in a natural way in all situations in which strategies are formed. Change becomes the 'normal' state of affairs and stability can merely be understood in terms of different forces balancing each other.

## Notes

1. Drafts of this chapter were presented at workshops and conferences in Berlin (July 1989), Leuven (September 1989) and Maastricht (November 1989). The author is especially grateful to Hein Schreuder, Arndt Sorge, Hans Pennings, Barbara Czarniawska-Joerges, Margaret Grieco, Jan Cobbenhagen, Michiel Roscam Abbing, John Hagedoorn, Dick Nelson, Jorge Katz and Henry Ergas for their constructive comments.
2. 'Formation', 'change' and 'transformation' are used here as interchangeable synonyms.
3. Scott (1987) correctly explains that one does not have to posit a survival need for the collectivity itself. It is sufficient to assume that some participants will have a vested interest in the survival of the organisation.

## References

Bartunek, M. (1984), 'Changing interpretive schemes and organisational restructuring: the example of a religious order', *Administrative Science Quarterly*, 29, pp. 355–72.

Burgelman, R.A. (1983), 'A model of the interaction of strategic behavior, corporate context and the concept of strategy', *Academy of Management Review*, 8, pp. 61–70.

Burgelman, R.A. (1985), 'Applying the methodology of grounded theorising in strategic management: a summary of recent findings and their implications', *Advances in Strategic Management*, 3, pp. 83–99.

Chaffee, E.E. (1985), 'Three models of strategy', *Academy of Management Review*, 10, pp. 89–98.

Clark, P. and K. Starkey (1988), *Organization Transitions and Innovation-Design*, London: Pinter.

Cohen, D., March, J.G. and J.P. Olsen (1972), 'A garbage can model of organizational choice', *Administrative Science Quarterly*, 17, pp. 1–25.

Cyert, R.M. and J.G. March (1963), *A Behavioral Theory of the Firm*, Englewoods Cliffs, NJ: Prentice-Hall.

Fiol, C.M. and M.A. Lyles (1985), 'Organizational learning', *Academy of Management Review*, 10, pp. 803–13.

Giddens, A. (1979), *Central Problems in Social Theory*, London: Macmillan.

Glaser, B.G. and A.L. Strauss (1967), *The Discovery of Grounded Theory*, Chicago: Aldine.

Grinyer, P.H. and J.C. Spender (1979), 'Recipes, crises, and adaptation in mature businesses', *International Studies of Management and Organization*, 9, 113–133.

Gross, N. and W.E. Martin (1952), 'On group cohesiveness', *American Journal of Sociology*, 57, pp. 546–54.

Harrison, J.R., Torres, D.L. and S. Kukalis (1988), 'The changing of the guard: turnover and structural change in the top-management positions', *Administrative Science Quarterly*, 33, pp. 211–232.

Hedberg, B.L.T. and S.A. Jönnson (1977), 'Strategy formulation as a discontinuous process', *International Studies of Management and Organization*, 7, pp. 89–109.

Hickson, D.J., Butler, R.J., Cray, D., Mallory, G.R. and D.C. Wilson (1986), *Top Decisions*, Oxford: Basil Blackwell.

Janis, I.L. and L. Mann (1977), *Decision Making*, New York: The Free Press.

Kanter, R.M. (1983), *The Change Masters*, New York: Simon & Schuster.

Maidique, M.A. (1980), 'Entrepreneurs, champions, and technological innovation', *Sloan Management Review*, 21 (2), pp. 59–76.

Masuch, M. (1985), 'Vicious circles in organizations', *Administrative Science Quarterly*, 30, pp. 14–33.

Maturana, H. and F. Varela (1980), *Autopoiesis and Cognition: the Realization of the Living*, London: Reidl.

Mintzberg, H. (1973), 'Strategy making in three modes', *California Management Review*, 16 (2), pp. 44–53.

Mintzberg, H. (1978), 'Patterns in strategy formation', *Management Science*, 24, pp. 934–48.

Mintzberg, H., Brunet, J.P. and J.A. Waters (1986), 'Does planning impede strategic thinking? Tracking the strategies of Air Canada from 1937 to 1976', *Advances in Strategic Management*, 4, pp. 3–41.

Mintzberg, H. and A. McHugh (1985), 'Strategy formation in an adhocracy', *Administrative Science Quarterly*, 30, pp. 160–97.

Mintzberg, H. and J.A. Waters (1982), 'Tracking strategy in an entrepreneurial firm', *Academy of Management Journal*, 25, pp. 465–99.

Mintzberg, H. and J.A. Waters (1984), 'Researching the formation of strategies: the history of Canadian Lady, 1939–1976', in Lamb, R. (ed), *Competitive Strategic Management*, Englewood Cliffs, NJ: Prentice-Hall.

Mintzberg, H. and J.A. Waters (1985), 'Of strategies, deliberate and emergent', *Strategic Management Journal*, 6, pp. 257–72.

Mintzberg, H., Ottis, S., Shamsie, J. and J.A. Waters (1986), 'Strategy of design: a study of "architects in co-partnership"', in Grant, J. (ed.), *Strategic Management Frontiers*, Greenwich: JAI.

Mohr, L.B. (1982), *Explaining Organizational Behavior*, San Francisco: Jossey-Bass.

Morgan, G. (1986), *Images of Organization*, Beverly Hills: Sage.

Nystrom, C., Hedberg, B.L.T. and Starbuck, W.H. (1976), 'Interacting processes as organization designs', in Kilmann, R., L.R. Pondy and D.P. Slevin (eds), *The Management of Organization Design*, I, pp. 209–30, New York: Elsevier North-Holland.

Penrose, E. (1959), *The Theory of the Growth of the Firm*, Oxford: Basil Blackwell.

Pettigrew, M. (1985), *The Awakening Giant: Continuity and Change in Imperial Chemical Industries*, Oxford: Basil Blackwell.

Robbins, S.P. (1984), *Essentials of Organizational Behavior*, Englewood Cliffs, NJ: Prentice-Hall.

Romme, A.G.L. (1990a), 'Vertical integration as organizational strategy formation', *Organization Studies*, 11 (forthcoming).

Romme, A.G.L. (1990b), Strategy formation in organisations. Dissertation, University of Limburg, forthcoming.

Romme, G., P. Kunst, H. Schreuder and J. Spangenberg (1990), 'Assessing the process and content of strategy in different organizations', *Scandinavian Journal of Management*, 11 (forthcoming).

Rothwell, R., C. Freeman, A. Horsley, V.T.P. Jervis, A.B. Robertson and J. Townsend (1974), 'SAPPHO updated-project SAPPHO phase II', *Research Policy*, 3, pp. 258–91.

Schon, D.A. (1963), 'Champions for radical new inventions', *Harvard Business Review*, 41 (2), pp. 77–86.

Schwenk, C.R. (1984), 'Cognitive simplification processes in strategic decision-making', *Strategic Management Journal*, 5, pp. 111–128.

Scott, W.R. (1987), *Organizations: Rational, Natural and Open Systems*, second edition, Englewood Cliffs NJ: Prentice-Hall.

Simmel, G. (1955), *Conflict and the Web of Group-Affiliations*, New York: The Free Press.

Soeters, J.L. (1986), 'Excellent companies as social movements', *Journal of Management Studies*, 23, pp. 299–312.

Starbuck, W.H. (1983), 'Organizations as action generators', *American Sociological Review*, 48, pp. 91–102.

Starbuck, H. and B.L.T. Hedberg (1977), 'Saving an organization from a stagnating environment', in Thorelli, H.B. (ed.), *Strategy + structure = Performance*, Bloomington: Indiana University, pp. 249–58.

Starbuck, W.H., A. Greve and B.L.T. Hedberg (1978), 'Responding to crises', *Journal of Business Administration*, 9, pp. 111–37.

Weick, E. (1979), *The Social Psychology of Organizing*, second edition, Reading: Addison-Wesley.

Weick, K.E. (1987), 'Substitutes for strategy', in Teece, D.J., (ed.), *The Competitive Challenge*, Cambridge, Mass.: Ballinger, pp. 221–33.

Wilson, D.C. (1982), 'Electricity and resistance: a case study of innovation and politics', *Organization Studies*, 3, pp. 119–40.

# 3. Management of innovation in the processing industry: a theoretical framework

*Jan Cobbenhagen, Friso den Hertog and Guido Philips*

## Issues

Managers have been frustrated by the slow rate at which new projects move from research to commercialisation, particularly because cash commitment increases sharply as the project moves into the development phase. The risks inherent in the failure to carry out essential development activities efficiently are all too obvious. The answer to the question why the organisation of renewal processes (that seem to be progressing too slowly) has to be changed, is simple: innovation is a way to survive — not only a defensive (because the competition is also doing it) but also an offensive way (to gain a competitive advantage). It is therefore of the utmost importance for a company to be able to innovate as effectively as possible. The importance of innovations will become clear when we look at a number of recent developments.

### *Time*

Life-cycles of products and technologies are becoming shorter. Technology and market changes are imposing themselves more and more rapidly. Being the first in the market and being able to deliver quickly gives a competitive edge. This applies to two different trajectories in the organisational chain: on the one hand, the production trajectory, including suppliers; and on the other, the renewal trajectory. At the moment, time is one of the most important competitive weapons: time savings allow companies to shorten their economic cycle, to innovate more rapidly, and to produce, sell and distribute more efficiently (Stalk, 1988; Uttal, 1987).

In fact, the main defence against the imitation of innovations is to keep ahead by creating a substantial lead time, which can be achieved through a continuous, incremental innovation policy. In most cases (except in the pharmaceutical industry, for example) patents provide only limited protection. Another defence is needed then, which can be provided, for instance, by the build-up of a knowledge base around the process, as can be found in the speciality industry. In this sector, process know-how is at least as important as the production equipment. Buying new technology is no guarantee of success for the newcomer. Process know-how is much more difficult to acquire; it takes time to develop it.

### *Product differentiation*

Particularly in the specialities and commodity-plus industries, it is important to be able to meet the demands of the customers as far as possible. Customising forces companies to produce smaller batches of a growing number of differentiated products. Products will become more knowledge-intensive. The increasing number of product variations one has to produce puts higher demands on the flexibility of engineering, production and logistics. In addition, markets are becoming more and more complex. The product itself is less important to the customer than its function. Thus, competition does not just include producers of the same product but also companies producing products that serve the same function.

*Heavy demands from market and environment*

The emphasis in organisational policy is shifting from the output (the primary process) to the innovation process. In this process, the environment confronts the organisation with conflicting demands: pace, creativity and flexibility on the one hand; and formalisation, standardisation and regulations (quality, liability, environmental protection and safety) on the other. A dilemma occurs here: the increasing competition confirms the need to innovate, but at the same time makes it more difficult for innovative products to be successful. To this end, concerted action is essential between core functions (e.g. marketing, R&D, mechanisation and manufacturing) at the business unit level. This will require an integral approach and draws attention to the challenge of overcoming the barriers between functional areas.

## Focus on the processing industry

This chapter is based on an investigatory study carried out for MERIT's research programme 'Management of Innovation in the Processing industry' (MIP). This programme will focus on the processing industry in general, and the chemical industry in particular. Innovation is inherent in the chemical industry, not only in its own field but also in other fields that are affected by its innovations. For example, as the production of its new compounds, products and materials integrate or replace natural ones, different processes and utilisation methods are required from those already known (Colombo, 1980). The developments mentioned in the previous paragraph are clearly noticeable in the processing industry as well. However, in this industry, and particularly in the chemical industry, some trends are probably more apparent than in other industries. In the first place, personnel management plays a critical role in both production and the innovation process. In production one grows more and more dependent on a decreasing number of people. As a result, a chemical plant can only be managed as a 'high commitment organisation': minor mistakes can have enormous consequences. In the innovation trajectory, we see that the ability to innovate is more constrained by the difficulty of finding the right people than by the money available for investments.

Furthermore, a qualitative change can be observed in the nature of innovations, albeit to a different extent in the various sectors. The differences between the sectors reveal themselves in the strategic attention concerning the process, the product technology and the extent to which they are 'market-driven'. In the speciality sector the conflicting demands mentioned in the previous section come to the fore in the innovation trajectory. In the commodity sector, however, they become evident in the decreasing control margins. The heavy chemicals and petrochemical industry has restructured its activites ever since the oil crisis in the 1970s. A shake-out period was followed by reorganisations, mergers and take-overs. Major factors contributing to the eroding profit margins were high labour costs, under-utilised capacity, heavy competition, prices under pressure, high energy costs and costs attached to the necessity of conforming with an increasing amount of government regulations. After years of depression this sector is now (1989) flourishing. The structural under-capacity makes it possible to keep profits at a high level. However, the commodity market will always be very sensitive to cyclical changes. The profits of a product can collapse drastically whenever a new plant is built. The depression in the commodity industry was one of the reasons why the processing industry started to diversify into specialties. In these high service-intensive speciality markets, products are sold on the basis of their performance rather than their composition or physical properties. That is why the opportunities in this market depend on the way in which one is able to respond to the dynamics of innovations and technological developments. The initially high profitability of this sector had the usual effect: everyone wanted 'a piece of the cake'. This led to an intense competition on the more popular speciality markets. The increased competition made it difficult for many speciality chemical companies to maintain their former levels of profitability. Some weaker companies could not survive or were taken over by stronger ones. Eventually, the markets matured and sales increased at a more normal rate. Despite the fact that the markets were mature, most firms were able to maintain high profit margins by providing a high level of field and technical service. In the 1980s the commercialisation of new technologies (such as microelectronics, genetic engineering) and certain high performance materials stimulated the growth of speciality suppliers who were able to differentiate themselves by superior product performance and innovation along with providing the necessary technical and field service support. A continuing desire to achieve higher profits is taking chemical companies further downstream beyond traditional products and markets into new high-tech businesses such as electronic chemicals and biotechnology (Boccone, 1988).

The differences between the various sectors make it necessary to break up the broad concept of innovation into smaller parts. We will distinguish here between product and process innovations. Product innovations can be sub-divided into product innovations where new developments in the technology play the most important part, and those where developments in the market are the major driving force.

Figure 1 shows the degree of importance of the various types of innovation for four sectors in the processing industry. Innovations in the mature and stable environment of the commodity industry are oriented towards the improvement of processes and the search for new applications of existing products, rather than the development of new products and processes. The commodity-plus sector is

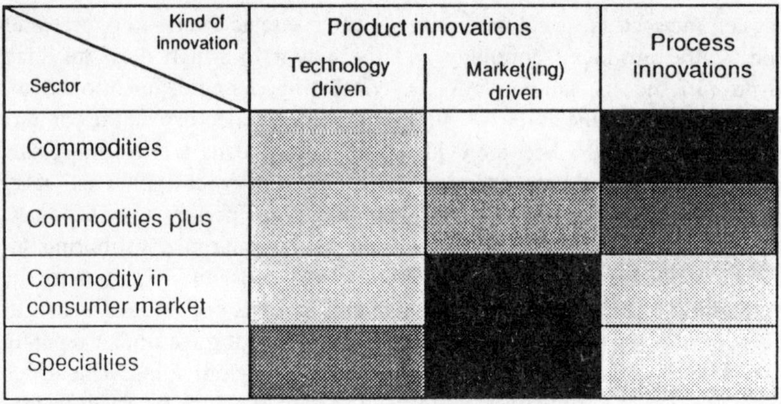

| Sector \ Kind of innovation | Product innovations | | Process innovations |
|---|---|---|---|
| | Technology driven | Market(ing) driven | |
| Commodities | | | |
| Commodities plus | | | |
| Commodity in consumer market | | | |
| Specialties | | | |

**Figure 1** Importance of different types of innovation for four processing industry sectors

more customer-oriented. This sector produces goods whose specifications can be modified to some extent to meet any special needs of customers (for example, ABS). Apart from process innovations, a number of product innovations are important in this branch as well. While innovations in this sector are driven particularly by technological developments, innovation in the commodity industry, which operates in the consumer market, is mainly market- and marketing-driven. This 'commodity in consumer goods' includes, among other things, detergents, foods like yoghurt, etc. Although product innovations are extremely important in these markets, they merely represent small adjustments to the physical product or modifications of the 'marketing appearance'. Contrary to the commodity market, where products of different manufacturers have the same chemical composition, the chemical composition of products in the specialities market vary from one manufacturer to another. This has a major impact on the performance of the product and makes it difficult for users to interchange specialities from different suppliers. At the same time, however, it provides opportunities for market-driven product innovations. Therefore it is not surprising that the speciality sector is dominated by the market- and marketing-driven product innovations, and that innovations in this sector are often initiated by customers. Being able to innovate fast is very important in this sector, since the product life-cycle of individual speciality chemicals is quite short and is decreasing even further. However, for most speciality chemicals, the period in which the product is profitable is substantially shorter than the life-span of the product itself. An active innovation strategy is therefore required. It is important that one is able to differentiate the products from those of the competitor. And as a result of the market-driven or sometimes even customer-driven characteristics of the new products we might expect that products are becoming increasingly differentiated.

Owing to this product differentiation and resulting pricing freedom, speciality chemicals usually generate substantially greater gross margins than do commodity chemicals. Unfortunately, the higher research and development and marketing expenses required in the speciality chemicals business erode much of these margins.

Although the focus in this chapter will be on product innovations, we stress the importance of both product and process innovations. In fact, process innovations are quite important in many of the speciality segments in the chemical industry. In some areas, such as those that apply biotechnology, process technology and process R&D innovations may become even more important in the future.

The above developments do not only lead to product and process innovations but to organisational renewal as well. Cumbersome formal organisational structures, often accompanied by informal structures which function well, will have to disappear and be replaced by more flexible and decisive structures. Organisational renewal is necessary because innovations bring with them a great deal of insecurity and risks, thus calling for different control techniques than the ones used within the functional areas. These control techniques are well in place in a stable and predictable environment, but they are inappropriate for the innovation process.

Innovations require that organisational demarcations and patterns be broken apart. Only then can integration take place. Therefore organisations need to change in order to be able to deal with an important innovation dilemma: control versus creative freedom. This dilemma is becoming most apparent in the area of tension between regulation and pace of innovation.

## Current organisational theory

A great deal of research is focused on the economic and organisational aspects of technology. Most of these studies, however, were characterised by a very segmentalist view. Organisational literature provides many approaches that are supposed to contribute to speeding up the innovation process. Typically, one of the oldest studies (Burns and Stalker, 1961) is still regarded as some kind of a fountainhead (Lammers, 1986). The main theme to be deduced from recent work of, for example, Mintzberg (1983), Moss Kanter (1983) and the success story of Peters and Waterman (1982) is the striving for organic organisation structures, propagated by Burns and Stalker. Mechanistic models, characterised by rigidity, co-ordination by regulations, individual responsibilities and a centralised decision process, have to give way to organic models. The latter, characterised by flexibility, co-ordination by consultation, collective responsibilities and decentralised decision processes, seem to foster extremely well in the dynamic and complex environments of innovations. Thus, innovation calls for an organic approach, a conquest of the bureaucratic segregation of functions, and it cannot function with formal communication patterns and a steep hierarchy. In addition to this main theme a number of sub-themes can be distilled from the literature.

### *Integrated versus segmentalist problem-solving*

An integrative view enables organisations to look at problems in their wider context. Problems will be considered from an integrated angle and the search for solutions will take place in co-operation with other departments. This is a must for innovations. Multifunctional product development teams (Whitney, 1988), also called 'rugby teams' (Takeuchi and Nonaka, 1986) or 'skunkworks' (Quinn, 1985), are a good example of this. Organisations with a more segmentalist culture seem to have more difficulty innovating. They hold on too long to their — in the past proven successful — way of handling things and will solve problems in isolation (Foster, 1987 and Moss Kanter, 1983).

*Business-orientation and entrepreneurship*

In addition to an integrative culture, another important precondition for success is a general business orientation. This orientation enables organisations to spread the control problem at system level over all other functions. In such a climate employees concentrate not only on the objectives of their own tasks but also on the developments and objectives at business unit level. Entrepreneurship is often regarded as a crucial factor for success for the innovativeness of small companies. Large organisations sometimes try to simulate such an entrepreneurial climate by appointing product managers or stimulating skunkworks (Quinn, 1985).

*Iterative versus linear*

Innovation processes are significantly more complicated than the linear models suggest. The disorderly and complex innovation process cannot be reduced to simple sequential (linear) models (Kline, 1985; Schroeder, 1986). Innovations which are organised on the basis of such models will progress too slowly and generally lead to less optimal products. They should on the other hand follow an interactive and iterative process allowing for feedbacks and alternative routes.

*Organisational structures*

Innovations call for an organisational structure that enables integrative actions. This does not mean strictly separated departments and management layers, but frequent interactions between departments and flat structures, in which teamwork and co-operation between disciplines is stimulated. Functional segregation has to be discouraged as much as possible (Moss Kanter, 1983).

*Human resources management (HRM)*

HRM is regarded as a strategic tool with the aim of supporting the innovation process. It is people who innovate, not machines; while short-term productivity can be influenced by purely mechanical systems, innovations require intellectual efforts of people (Moss Kanter, 1983). Therefore, people are an investment and personnel turnover a loss. The 'chaos' theory tells us that organisations have to be 'self-organising' (Prigogine and Stengers, 1984). Managers will have to navigate rather than control, and they should allow their subordinates to take risks and make mistakes. Innovations sometimes require decisions regarding problems for which no routine approach is available. To this end subordinates ought to be given freedom and opportunities to develop new approaches.

*The creative chaos*

Strongly structured planning systems generate a relatively small amount of important innovations (Quinn, 1985). Innovations appear to do well in an environment that can be characterised as a creative chaos. In fact, Peters's ideas (1988) merely provide an illustration of this theory. They do not add any substantial new elements. But even the basic concepts of the theory of chaos are not as new as most people think. In 1959 Dalton already stated that changes towards new directions in general and creativity in particular can be only partially subject to rules: 'With its larger blessings participative organisation embraces some disorder.'

Above we referred to some 'recipes' for innovative success that can be found in the literature. The practical applicability of these recipes, however, is restricted because each of them focuses on just part of the innovation dilemma. They merely offer a basis to start from. Loose approaches are moreover far from optimal: a certain recipe may lead to the intended effect but, due to its isolation, the effect will not last for long. In fact it may even conflict with the original objectives. Sometimes the recipes for several partial problems will clash with each other, thus leading to an unpredictable and rarely effective result.

The risk of researchers approaching the innovation phenomenon in a segmentalist way is similar to the risk of managers whose efforts to manage innovations are guided by a few classical basic perceptions. Figure 2 illustrates the pitfalls that can arise when using a number of basic assumptions which

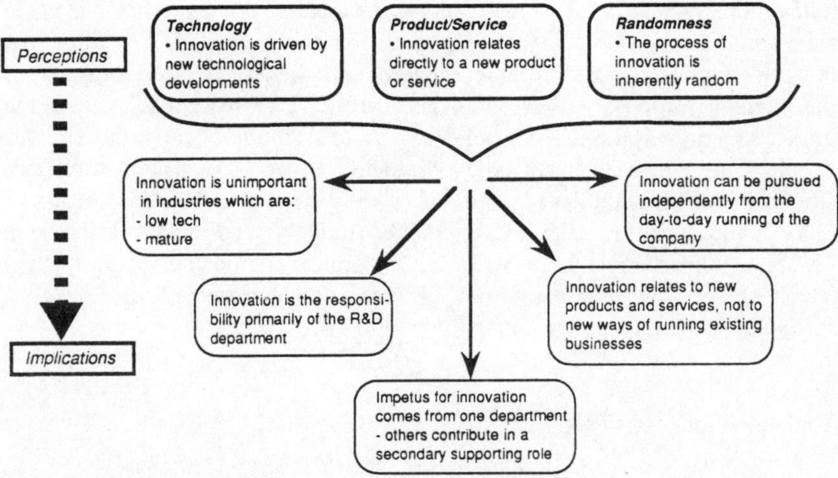


**Figure 2** Basic perceptions determine how companies approach innovation

*Source:* O'Hare, 1988, p. 22

O'Hare (1988) observed in practice. Assuming that definitions determine perceptions and perceptions drive action, O'Hare has listed some classical basic perceptions managers tend to use with regard to innovations. The implications of these perceptions tell us that the possibilities a manager considers regarding the innovation problem are restricted to a great extent by the definition of innovation he or she uses. Here, we see very clearly that a segmentalist view can have far-reaching consequences. For example, if the manager thinks that the innovation process is a randomly occurring process, which is closely linked to technology and new products or services (as Figure 2 illustrates), we see that these perceptions have implications which are quite understandable from the manager's point of view. But when viewed from a more integrated angle, they represent an inappropriate innovation policy.

The above-mentioned segmentalist approach found in most literature shows that the studies have focused on just parts of the innovation dilemma. But not only has the organisational research itself been segmentalist, it has even failed to integrate with the economic tradition. The organisational tradition has been mostly concerned with the impact of technology on firm members and the social context of change. A large number of researchers have neglected the economic aspects of technological change. As Grindley (1989) argues, organisational researchers (Burns and Stalker; Woodward; Trist; Child; *et al.*) have primarily studied the internal implementations of change without linking these to external competitive necessities. Both the organisational tradition and the economic tradition have remained poorly integrated. In its turn, the economic tradition (Griliches, Kamien and Schwartz; Mansfield; *et al.*) has been concerned mainly with technology as being part of market behaviour between firms and has stopped well short of the organisational context of change.

Some literature looks upon technological change as being abrupt instances of change. Recent work has shown that innovation and technological changes must be seen as a continuous process. Therefore, we need a process theory of innovation. Such a theory is not available as yet, although the MIRP has made some progressive steps in the right direction (Schroeder *et al.*, 1986). An integrative vision is called for here: separate approaches have to be bound into one integrative approach. Such an integration often leads to a rough image of the innovating organisation, some kind of '*Gestalt*'. In the ideal situation, this should be a consistent 'ideal type', characterised by an internally consistent structure. All the fragmented approaches will be integrated in one consistent unity. Some efforts have already been made. Mintzberg's adhocracy (Mintzberg, 1983) is a good example. In his vision, an integrative approach is an approach in which there is a perfect internal fit between sub-systems, control tools, structures, personnel policy and culture.

## The search for consistency

If cognitive psychology has made one thing clear, it is that in human thinking we can detect a continuous striving for consonance. Unconsciously as well as consciously we want our image of the world in which we are functioning to correspond with reality. Whenever we receive new information that contradicts our notions, our knowledge or our behavioural patterns, some thinking processes will be triggered whose aim is to make our image fitting again. Or, put in the terminology of cognitive psychologists: to be consonant again. This process is better known as 'dissonance reduction'. Dissonance reduction can occur by ignoring new information, transforming old information, changing our behaviour or adapting our emotions. This process can be either purely rational or irrational. For instance, take the reaction to a supplier who has been delivering on time for a couple of years, and suddenly comes a few days late. A rational reaction is to check what is going on and perhaps to set new conditions. However, our reaction can be classified as irrational when we immediately turn to his competitor. For we simply forget his excellent delivery record in the past and the fine relationship. We erase relevant information as it were.

A similar phenomenon can be observed in organisational literature. Organisations are effective when the elements of their structure and culture are mutually consistent and when they correspond to their environment. Henry Mintzberg, one of the most prominent representatives of modern organisational theory, expresses the essence of this phenomenon in one sentence: 'The elements of structure should be selected to achieve an internal consistency or harmony, as well as a basic consistency with the organisation's situation — its size, its age, the kind of environment in which it functions, the technical systems it uses, and so on' (Mintzberg, 1983, pp. 2–3). This argumentation of the organisational fit almost seems like forcing an open door. Who would dare to claim the opposite: who will look for inconsistency and who would not try to adapt to the circumstances?

It is not our intention here to plead for the abolition of the above-mentioned central paradigm in organisational science. Its usefulness in organisational practice and organisational research is too distinct. We do not wish to attack the paradigm of consistency by switching to the opposite or replacing it by a paradigm of inconsistency. We merely want to make clear that inconsistencies are inherent in innovations; inconsistencies are both a source as well as an effect of innovations. Therefore innovative success is largely based on handling inconsistencies effectively. This implies that in some cases one must eliminate inconsistencies, while in other cases it is legitimate to strive for them (or even to provoke them).

Essentially, inconsistency is the source from which successful renewals spring. In addition, renewal is in itself in most cases a source of inconsistency. Inconsistency is inherent in innovation. This reveals itself most clearly in the destructive nature of innovation. Innovation does not just provide a basis for an increase in productivity, or the exploration of new markets; it is not just a challenge to the organisation. Innovation also swallows money. Wrong choices may endanger the continued existence of the organisation. New products can jeopardise existing products that bring in the cash. This threat can be so huge that the innovation is postponed with the underlying reasoning: 'Why should we cannibalise our profitable business for uncertain profits from a rapidly changing solid state business? It makes no sense to cannibalise a proven source of income' (Foster, 1987). It is not unusual that 80 per cent of the technical expenses are allocated to the defence of products that are considered more important because of their contribution in the past couple of years, rather than on the basis of their future contribution, as a survey of McKinsey showed (Foster, 1987). This happens, in spite of the fact that one is well aware of the long-term risks involved in postponing the investment in a new technology.

Innovations do have direct consequences for the status quo within organisations. Every technological development that breaks through at the expense of others confronts the organisation with personnel problems. Take, for example, the position of analogue programmers: a large number of them became problem cases for the personnel department in their mid-thirties. The problem of the older development engineer manifests itself at an increasingly early age.

In the literature we find some references to the concept of inconsistencies, two of which will be briefly discussed here. Firstly, Jay Galbraith (1984) simplifies the success of product innovation as both the creative ability as well as the ability to place one's self in the position of potential users and others who have to work on the new product in the development chain. This points in the direction of greater independence for researchers and developers. Within the organisation a hedge is place around the new product and the project group involved. But, on the other hand, innovation also requires a confrontation; a confrontation with the market and all the people that are somehow involved further in the innovation chain: from process engineering to production and distribution. 'We won't get anything finished if we have to work like that' will be the initial response of a developer. But despite this comment, both elements occupy a central position in modern management literature on innovation.

The second inconsistency relates to the control of the innovation process. Innovation involves considerable risks. It is not surprising that the process of innovation is sometimes referred to as 'the harnessing of the beast'. Innovating successfully calls for a creative chaos, preferably one that can be controlled (Quinn, 1985). However, the process of innovation cannot be controlled to the extent desired. Top management ought to be more involved and projects need better planning techniques. On the other hand, classical linear planning and control systems no longer seem to be satisfactory, and innovation teams have to be freed from their grip. Clues and other indications telling us how to deal with these inconsistencies are very hard to find.

## New leads in innovation theory

There are still certain limitations attached to the theories and models that we find in modern management literature. On the one hand we find descriptive models which merely answer the question 'why are we the way we are?' The manager who learns from a researcher that his organisation has the characteristics of an adhocracy and that this can be easily explained, will in most cases merely take note of this announcement, and just think: 'So what. . .?' Normative ideas and models, on the other hand, provide a direction towards which an organisation has to proceed in order to innovate successfully. The answer to how the organisation starts to move and how management can guide these movements has received very little attention as yet. Which methods, strategies and instruments are available? How can the shift from A to B be translated into a concrete policy? These are questions to which the current management literature has not yet found a satisfying answer.

One of the risks attached to the development of theories is that these often reflect the complexity and ambiguity of the phenomena they try to explain. This is certainly a potential threat in the area of innovation. In addition, the users of new insights prefer simple metaphors when dealing with complex organisational problems. In the remainder of this chapter, we will not attempt to produce all-embracing models. Instead, we will restrict ourselves to two basic ideas which can be elaborated in later stages. These basic ideas can be expressed as propositions:

1. Innovation management depends on the way in which one is able to deal with inconsistencies.
2. The innovative capability is built in a learning organisation. Therefore, the control of innovation processes calls for a shift from formal planning (feed forward) towards an investment in feedback.

This approach fits well in the development of a process theory of innovation. In such a theory, innovation is regarded as a stream of events across the various columns of the organisation. This process theory may serve as a medium to bundle concrete experiences in companies.

*Management of inconsistencies*

Although 'inconsistencies' lead to numerous negative effects (personnel and economic frustrations, turnover of malcontents, dragging of feet), they also have important positive effects. In fact, they can be depicted as 'fluctuating around a fragile balance of disruption *and* construction'. However, people are reluctant to provoke them in order to achieve something constructive.

In stating that the management of innovation is in fact 'dealing with inconsistencies in an effective way', we do not mean that one should refrain from attending to apparent inconsistencies that can be solved. Nor does it mean that one should start an active search for inconsistencies and conflicts. What we do mean is that everyone at all levels in the organisation should be convinced that inconsistencies and conflicts are inherent in the innovation processes and require specific attention. However, as Dalton (1959) points out, this does not mean that inconsistencies and conflicts will always be accompanied by personnel creativity or general progress. But conflict and co-operation are usually intermingled in all advances, especially in democracies.

Let us give an example. A project seems to get stuck within the tribal battle of functional departments. Top management therefore decides to have the project carried out in a new project organisation in another building across town. The project gets its own staff, budget and a direct line to the top. Despite the fact that in the current situation this measure will perhaps be the only one that offers some prospect of a good product, one eventually has to face some consequences. At

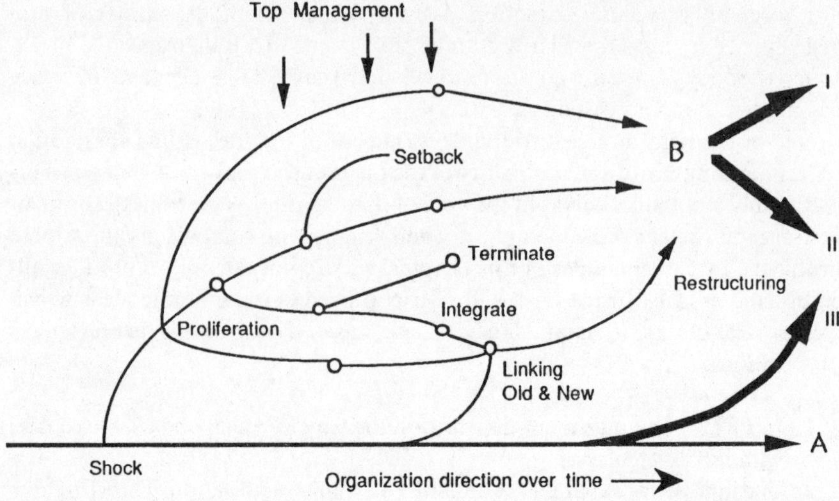


**Figure 3** Illustration of the innovation process model by the MIRP

*Source:* Schroeder *et al.*, 1986, p. 519

some point in time the functional departments of the mother organisation have to be called in. And at some point in time, the product has to be incorporated in the existing assortment and in the 'normal' management. Also, there will be a moment when the staff has to return to the old organisation. The introduction of this organisational 'antipode' brings new inconsistencies which should not be neglected. These may have a harmful effect on the product in question, or otherwise on the following products. The challenge is to monitor the new and latent inconsistencies, in order to interfere whenever that is necessary.

The innovation process can be regarded as a chain of events that originates from a shock or a mutation in our thinking (see Figure 3). It will take a while until the person(s) that came up with the idea has gathered enough sponsors and the top will commit itself to a new innovation trajectory. This one strong idea will result in a process of proliferation. Numerous activities are set in motion which have to ensure the eventual ability to produce and commercialise. More and more persons and groups within the organisation will become involved. New materials have to be tested. The existing measuring instruments will probably be inadequate. One has to gain more insight into the needs and desires of the various customers. Quality has to be built in from the beginning, and the control department has to make extra effort to allocate and control the various costs. Proliferation not only causes the activities to become increasingly difficult to link to the baseline. Also, tensions between sub-trajectories will be occurring more frequently; for instance, when the group that is developing the new materials discovers a new perspective. However, holding on to this new line would significantly delay the project and render the work of a group of process technologists practically superfluous.

In other sub-trajectories there is a danger of being caught in vicious circles. As a result of this pressure, one does not usually bring the problems out into the open. A small circle may thus develop into a snowball until it becomes a fatal mistake.

In the process of proliferation one eventually enters the domain of other products: either the product no longer fits in the assortment or it becomes a direct competitor of another product which has a lot of costs sunk in it. In this phase the manager sighs: 'The problem is like trying to grow an oak tree when there are inexorable pressures to grow a bramble bush' (Schroeder *et al.*, 1986).

In our view, this proliferation process is inherent in the innovation process. Organisations differ to the extent to which they can control this process. These patterns emerge also in companies which are extremely successful. Furthermore, the character of product or process renewal itself plays an important part: as the uncertainties (the newness) and complexity (the involvement of the various disciplines and interest groups) grow larger and the through-put time longer, the proliferation process threatens to become more intense.

What does this analysis mean for the management of innovation? In the first place, it is essential that the most important 'built-in' inconsistencies are recognised by business management; not with the intention of solving all the problems in advance, but to be able to fathom the situation regularly. This is referred to as 'management of attention'. Management has to recognise the critical factors that can pace and delay the process. These factors should receive their regular attention, so that management can intervene whenever necessary. One could call this the confrontation phase: a confrontation between the initial goals with the present state of the art. But also a confrontation between what is thought of upstream and what one has to work with downstream. And finally, a confrontation with the market; with the concrete demands of the customer and the progress made by the competition.

In order to deal with inconsistencies, the following range of interventions can be chosen from:

*Stopping an activity on time.* This takes courage, because it means a loss of face for sponsors and the people who came up with the ideas.

*Isolating an activity within the project.* An essential activity is hedged and protected in order to prevent other activities from interfering with it.

*Isolating an activity outside the project.* One discovers that tests with new materials offer new possibilities, but at the same time one finds out that the product development is slowed down considerably. The project is then taken out of the main stream and will be used for later products.

*Integrating activities.* One recognises that the same problem is dealt with in three separate projects, for different products, yet with the same purpose.

*Buying know-how.* It takes too long, or it is far too expensive, to come up with one's own solutions, so know-how is bought from others.

*Accelerating and supporting successful activities.* Some activities seem to attract synergy. Backing these activities will have a stimulating effect on the involvement of others.

*Adjusting activities.* Objectives of sub-projects are reformulated and adapted to each other.

*Selling or disposing one's self of some or all the development tasks to other companies.*

The progress of the innovation process depends on the way in which project leaders and project members from all kinds of functional departments and (sub-)project groups co-operate. This is where problems are made and solved. Located between this group and business management we will find, in most organisations, the layer of functional management, which is responsible for the functioning of certain experts and keeping up their expertise. In actual practice, however, this layer often works like a 'clay layer'. Top management is constantly called in to solve interface problems at primary level. In the 'management of inconsistencies' these roles are reversed. The business management's task is to interfere at critical moments and critical places and to ask those involved how *they* think they can solve the problem.

One of the reasons why bureaucracies and mechanical organisations are not usually innovative is because they strive to cast out the inconsistencies. Bureaucracies in particular, rigorously try to eliminate inconsistencies. This may, of course, eliminate uncertainty and some moral suffering, but, from an innovative point of view, the disadvantages outnumber the advantages.

### *The learning organisation*

A competitive advantage can be achieved through a process of continuous renewal. Continuous renewal (innovation) is basically a learning process of gathering knowledge, transferring knowledge, storing knowledge and applying knowledge. It is not surprising that the literature attributes a central place to the learning ability of organisations. 'Organisational learning', defined as 'the process in which an organisation makes use of information from past events to master future events better' (den Hertog and van Diepen, 1988) is vital to the successful management of innovations. This learning process cannot be seen as an efficiency gain by routinising and indefinitely repeating operations, as is the case in classical mass-production. On the contrary, the learning process is found on a much higher level within the innovating organisation. The economic discussion on learning curves, which focuses on mass production, distinguishes between embodied and non-embodied learning effects. Embodied learning effects represent the collection of all the knowledge and skills that is stored in machines, product designs and patents. Non-embodied learning effects are formed in the minds of people. The use of words here is almost symbolic of the implicit outlook on the role of people, groups and organisations in the learning process: is it not more logical to speak of 'embodied knowledge' when we mean human knowledge?

The value attached to embodied and non-embodied knowledge varies substantially from company to company. On the one hand, there are companies which — upstream and close to the acquisition of knowledge — steer towards reinforcement of their non-embodied knowledge. They are convinced that personnel is not a cost item, but an investment (asset). The smartest companies moreover pay much attention to the transfer and use of knowledge: from research to development, from mechanisation to the plant and vice versa.

On the other hand, there are companies which invest hundreds of millions in new product development with the purpose of reinforcing only the embodied knowledge. Their fixation on the final product is extreme to the extent that at the time when the large research and development teams are dissolved no-one at policy level wonders what to do with the knowledge and experience gathered. When replacing a development group, very often only the out-of-pocket costs are considered, though rarely the consequences for the knowledge base.

Contrary to what is usually thought, this is not a matter for R&D alone. The learning processes within the plant, marketing and especially those between the functional departments are of equal importance. The question that arises is how to raise the learning capacity of (parts of) an organisation. The learning capacity is defined here as: 'the capacity to search for information, to integrate, to store and retrieve, to transfer and to apply it' (den Hertog and van Diepen, 1988). The question is whether the learning capacity can be influenced and increased. Our proposition is that it can. Below we will briefly discuss four ways to achieve this:

*Organisational design*

The control capacity and thus the learning capacity, is determined to a large extent by the structure of the organisation. In order to overcome functional segregation within the organisation, it is important to use optimal control capacity at the lowest possible level within the organisation. Business units, operational groups and autonomous task groups represent better structural conditions for the learning capacity than organisational structures dominated by functional segregation. Skunkworks or reservations, as Galbraith (1984) calls them, serve as 'havens for safe learning'. In these sub-organisations, which are totally devoted to creating new ideas for future business, it is possible to maximise early failure to promote learning. In order to be able to learn, one has to see the connection between cause and effect. Organisational structures have to provide the means to achieve this.

*Personnel policy*

Personnel policy is, particularly upstream, a powerful tool to reinforce the learning capacity. It is important that it is made explicit in personnel policy to monitor the flows in which individuals and groups move, and in which projects, departments, expert fields and responsibilities succeed each other. In companies we often see regular patterns and segregations which are hard to break. Young academics nearly always flow downstream, from research to development, engineering, production or marketing, while recent research shows that upstream movements may bring with them a significant added value to the organisation. This is just one example. We could give numerous examples which show that the non-embodied knowledge can be increased by an integrated use of the classical instruments of personnel policy (recruitment, selection, career planning, training and rewarding).

*Feed-forward versus feedback*

The uncertainties and inconsistencies caused by innovations call for different control techniques than the usual feed-forward planning. The detailed planning and decision methods of systems management, placing a strong emphasis on planning and feed-forward, have long served as control tools for innovation. Nevertheless, successful innovation appears to be stimulated particularly by efficient feedback. Feedback control techniques allow the organisation to learn from experiences. For that reason it is important that a culture be developed in which the lessons learned from mistakes and successes are systematically recorded. Learning experiences have to be documented well and less attention has to be given to the formulation of formal plans in advance. This mainly concerns the documentation of casual relationships: do we know why we always have to deal with problem *x*, or why certain actions always succeed others? These findings are then evaluated and fed back. In this way, the lessons to be learned from one innovation process can be of use for a next one; the implications that actions and experiences in a certain phase can have for previous phases become distinct and can be dealt with in later product renewal projects.

However, such a planning technique is rarely used in practice yet. 'Learning', and particularly 'organisational learning', does not represent an explicit goal of innovation processes. Experiences and mistakes are rarely documented and evaluated, so that the same mistakes are made over and over again. Therefore, documenting and evaluating (feedback planning) are important tools for innovation management (Cobbenhagen, 1990). Management on the basis of feedback, though, will nearly always involve a cultural change. Not only because we were brought up with a great sense of responsibility but also because we live in a culture that imposes negative sanctions on bad news and messages which conflict with the existing values and norms. And above all because our culture reflects an undervaluation of the non-embodied knowledge.

*Experimenting*

Particularly automation projects tell us how vulnerable linear trajectories are. As soon as the integrated system finally runs, it often becomes obvious that the original assumptions are outdated. However, this trajectory can be accelerated by artificially creating situations in which failures and deviations that pose real threats are simulated. In this way the organisation as a whole can learn how to deal with similar situations and the preliminary design can be adapted to handle them in real life. Prototyping and experimental settings are more useful when they are placed within the 'real organisational' context. Otherwise, only the experts who are designing the system can learn from these experiments, and not the final users.

Organisational learning is especially a matter of breaking old habits and behavioural patterns that are no longer appropriate. That is why preaching it is a lot easier than actually putting it into practice.

## Mission for the 1990s

The processing industry, and particularly the chemical industry, is considered as a modern sector of industrial activity. The fully computerised plant with its modern laboratories and tinted windows represents the 'ideal type' company of the future, like the machine and assembly plant used to in the past. The chemical and, especially, the speciality branches are constantly looking for new types of work and organisation which go beyond the present organisational models. A number of companies, especially young organisations which are at the 'top end' of technological development, can boast of an excellent state of organisational renewal. Just because of this it is highly important for organisational scientists to pay attention to this environment.

The chemical industry in particular is flourishing at present (1989). In a country like the Netherlands its contribution to export is highest after agriculture. This development is not free of risk. Just go back in time ten years: at that time the commodity sector of the chemical industry sounded the alarm, and everybody wondered whether there was a future for this branch in Europe. After a thorough reorganisation, which led, among other things, to a slight under-capacity in a large number of markets, this branch of the chemical industry was given a 'clean bill of health' again. Meanwhile, however, the nature of the organisational problems in this commodity sector, which is at the 'low end' of technological development, has become very clear. A large number of oil refineries, salt and food processing industries are presently facing problems that are inherent in the classical plant organisation. Remarkably, this is where we can learn a lot about organisational renewal in discrete batch production, particularly when dealing with integration of line and staff responsibilities. On the other hand, as a result of the integration of processes and parts, discrete batch production seems to conform more and more to the process type of production. Thus, although this chapter focuses on the management of innovation in 'the processing industry', many of the conclusions that are drawn and the suggestions for best practice can be regarded as suggestions for best practice in the management of innovation generally rather than necessarily being specifically tied to the process industry.

When things are going well and a great deal of money is made, we are tempted to postpone difficult issues until tomorrow and concentrate on making money today. A lot of companies at the top end of the processing industry, where the focus is on R&D rather than production, seem to be able to indulge in such a luxury. However, they tend to forget about the risks of tomorrow involved in their preoccupation with today's success. In the first place, these risks stem from the environment, which makes increasingly stiff demands on the activities carried out within the organisation, e.g. demands in terms of environmental protection, safety, quality and health. Secondly, fossilisation and obsolescence will inevitably take effect as time goes by, even in new organisational structures. At the moment when there is no direct economic need to do so, companies will have to invest not only in new products and processes, but in their own organisation as well. This applies particularly to the new companies established within the large, older undertakings. Smaller, young companies which have thrown themselves in the high-tech market seem to have fewer problems. And it is very likely that they owe their success to their unorthodox working method.

This is not an easy task for these companies dominated by technologists. Even more so since they are operating in an environment in which organisational theory offers so little to hold on to in terms of renewal. The 1990s seem to offer even less scope for economic renewal, and at the same time there will be an increasing need to adapt to the altered circumstances.

In this chapter, we postulated that innovations always create an area of tension between consistency and chaos, between stability and instability in the organisation. Since these inconsistencies are inherent in innovations, it is important that one is able to deal with them. As we have hopefully shown, the *management of inconsistencies* is the key to this process.

## References

Abernathy, W.J. and K.B. Clark (1985), 'Innovation: mapping the winds of creative destruction', *Research Policy*, 14, pp. 3–22.

Boccone, A. (1988), 'Profits evolve in specialties business', *ECN Specialty Chemicals Supplement*, January, pp. 5–11.

Burns, T. and G.M. Stalker (1961), *The Management of Innovation*, London: Tavistock.

Child, J. (1984), *Organisation: a Guide to Problems and Practice*, London: Harper & Row.

Cobbenhagen, J.W.C.M. (1990), 'Innoveren: strategieën en modellen', in Hertog, J.F. den, Eijnatten, F.M. van (eds), *Management van technologische vernieuwing*, Assen/ Maastricht: van Gorcum.

Colombo, U. (1980), 'A viewpoint on innovation and the chemical industry', *Research Policy*, 9, pp. 204–31.

Dalton, M. (1959), *Men who Manage*, New York: John Wiley.

Foster, R.N. (1987), *Innovation: the Attacker's Advantage*, London: Pan.

Frontini, G.F. and P.R. Richardson (1984), 'Design and demonstration: the key to industrial innovation', *Sloan Management Review*, Summer, pp. 39–49.

Galbraith, J.R. (1984), 'Human resource policies for the innovating organisation', in Fombrun, C., Tichy, N.M. and Devanna, M.A. (eds), *Strategic Human Resource Management*, New York: John Wiley.

Griliches, Z. (1958), 'Hybrid corn: an exploration of the economics of technological change', *Econometrica*, 25, pp. 501–22.

Grindley, P. (1989), 'Technological Change within the Firm: a Framework for Research and Management', paper presented at the 3rd annual conference of the British Academy of Management, Manchester, 10–12 September.

Grönlund, A. and S. Jönnson, (1989) 'Managing for Cost Improvement in Automated Production', paper presented at the colloquium on 'Measuring Manufacturing Performance' Harvard University, graduate school of business administration, 25–26 January.

Hertog, J.F. den and S.J.B. van Diepen (1988), 'Technological innovation and organisational learning', in Bullinger, H.J. (ed.), *Information Technology for Organizational Systems*, Amsterdam: North Holland.

Hirsh, R.F. (1986), 'How success short-circuits the future', *Harvard Business Review*, March–April, pp. 72–6.

Imai, K., Nonaka, I. and H. Takeuchi (1985), 'Managing the new product development process: how Japanese companies learn and unlearn' in Clark, K. *et al.*, *The Uneasy Alliance*, Boston: Harvard Business School press.

Kamien, M. and N. Schwartz (1982), *Market Structure and Innovation.* Cambridge: Cambridge University Press.

Kline, S.J. (1985), 'Innovation is not a linear process', *Research Management*, July–August, pp. 36–45.

Lammers, C.J. (1986), 'De excellente onderneming als organisatiemodel', *Harvard Holland Review*, 8, herfst, pp. 18–27.

Leonard-Barton, D. and W.A. Kraus (1985), 'Implementing new technology', *Harvard Business Review*, November–December, pp. 102–10.

Maidique, M.A. and R.H. Hayes (1984), 'The art of high-technology management', *Sloan Management Review*, Winter, pp. 17–31.

Maidique, M.A. and B.J. Zirger (1985), 'The new product learning cycle', *Research Policy*, 14, pp. 299–313.

Mansfield, E. (1968), *Industrial Research and Technological Innovation*, New York: Norton.

Mathôt, G.B.M. (1982), 'How to get new products to market quicker', *Long Range Planning*, 15 (6), pp. 20–30.

Mintzberg, H. (1983), *Structures in Fives: Designing Effective Organizations*, Englewood Cliffs, NJ: Prentice-Hall.

Moss Kanter, R.M. (1983), *The Change Masters: Innovation for Productivity in the American Corporation*, New York: Simon & Schuster.

Quinn, J.B. (1985), 'Managing innovation: controlled chaos', *Harvard Business Review*, May–June, pp. 73–84.

O'Hare, M. (1988), *Innovate!*, Oxford: Basil Blackwell.

Peters, T.H. (1988), *Thriving on Chaos: Handbook for a Management Revolution*, London: Macmillan.

Peters, T.H. and R.H. Waterman (1982), *In Search of Excellence: Lessons from America's Best Run Companies*, New York: Harper & Row.

Prigogine, I. and I. Stengers (1984), *Order Out of Chaos: Man's New Dialogue with Nature*, London: Heinemann.

Schroeder, R., A.v.d. Ven, *et al.* (1986), 'Managing innovation and change processes: findings from the Minnesota innovation research program', *Agribusiness*, 2 (4), pp. 501–23.

Stalk, G. (1988), 'Time, the next source of competitive advantage', *Harvard Business Review*, July–August, pp. 41–51.

Takeuchi, H. and I. Nonaka (1986), 'The new new product development game', *Harvard Business Review*, January–February, pp. 137–46.

Trist, E. (1981), 'The evolution of sociotechnical systems as a conceptual framework and as an action research programme', in van de Ven, A.H. and W. Joyce (eds), *Perspectives on Organizational Design and Behaviour*, New York: Wiley.

Tushman, M.L. and W.L. Moore (1982), *Readings in the Management of Innovation*, Boston: Pitman.

Uttal, B. (1987), 'Speeding new ideas to market', *Fortune*, 115 (5), pp. 54–7.

Ven, A.H. van de (1986), 'Central problems in the management of innovation', *Management Science*, 32 (5), pp. 590–607.

Whitney, D.E. (1988), 'Produceren volgens ontwerp', *Harvard Holland Review*, 17, winter, pp. 41–9.

Woodward, J. (1965), *Industrial organization: Theory and Practice*, Oxford: Oxford University Press.

# 4. Technical innovation in the world chemical industry and changes of techno-economic paradigm

C. Freeman

## 1. Introduction

This chapter will discuss the changing conditions affecting innovation in the chemical industry from the 1930s to the present time and some prospects for the future. In Sections 2 and 3 the main focus is on innovation in synthetic materials and pesticides, which were among the fastest growing sectors of the industry from the 1930s to the 1970s. The success of the leading chemical firms over this period is related to those general factors affecting success and failure, which were identified in well-known studies of innovation, such as Project SAPPHO and the TRACES project.

Section 4 discusses the *discontinuities* in technology which became increasingly apparent in the 1970s and 1980s. These discontinuities are so important that they may be described as a change of 'techno-economic paradigm'. The shift from energy-intensive mass and flow production systems to information-intensive flexible production systems, based on microelectronics, affects all industries, including chemicals. The final section of the chapter discusses the extent to which the national environment or system of innovation facilitates this transformation in Japan and elsewhere, and the future prospects for world technological leadership.

The justification for a historical approach to the contemporary problems of the chemical industry is twofold: scientific and practical. On the one hand, the social sciences must use historical data to generate hypotheses, theorems, analogies and generalisations. It is, of course, essential in using such data to take into account discontinuities as well as continuous trends and phenomena. Historical research on technical and organisational innovation is especially important for economic and management sciences. It shows that there are identifiable patterns of ordered change in technology, as well as some unpredictable radical discontinuities (Dosi *et al.*, 1988). Only a historical perspective permits us to approach these problems.

The history of technical change over the last hundred years shows that the technology embodied in new chemical products and processes comes from many different sources. The art of innovation and of innovation management lies precisely in combining multiple, and often diverse, sources of information and ideas. The main sources of new technology may be classified as in Table 1.

This chapter will argue that, since the emergence of the German synthetic dyestuffs industry in the last quarter of the nineteenth century, the most successful and innovative firms have been characterised by a strong *in-house* capability for R&D, generating a corporate research tradition and firm-specific processes of technology accumulation. But the results of empirical research on innovation (e.g. National Science Foundation, 1973) conclusively demonstrate that this in-house technological strength is accompanied by many linkages to external sources of scientific and technical information and ideas, as well as to markets, user experience and production (points [2] and [3] in Table 1).

The particular ways of combining the various external sources of new knowledge and information (points [4] to [14] in Table 1) vary in different firms, industries and countries. Especially important in the chemical industry has been access to the results of basic research in universities (see, for example, Gibbons and Johnston, 1974; National Science Foundation 1973). The mode of access to these results has varied from simple recruitment of graduate chemists and engineers to the elaborate networks of consultancy and university research endowment developed by the German chemical industry. But the major innovations of the chemical industry could not have been made without the fundamental scientific base developed primarily in European universities (Freeman *et al.*, 1963).

But although contact with the results of fundamental research in universities was essential, in-house R&D was generally the most important source of new technology for European and US chemical firms from the 1930s to the 1960s. Whereas in their classic study, *The Sources of Invention*, Jewkes *et al.* (1958) argued that many important twentieth-century products were first invented in universities or in small firms and not in the R&D laboratories of large firms, they made an exception of the chemical industry.

For *incremental* innovations the experience of production, plant design and construction, and of marketing and user feedback, was the most important stimulus to new development in technology and the most important source of new ideas and information, i.e. (2), (3) and (4) in Table 1. But in-house R&D (1),

**Table 1** Sources of new technology

(1) In-house research, design and development
(2) Experience in production, quality control and testing
(3) Experience in marketing and feedback from users
(4) Experience in plant design and construction and feedback from contractors and suppliers
(5) Scanning of world scientific and technical literature, patents and other information sources
(6) Recruitment of engineers and scientists
(7) Contact with university science and engineering faculties
(8) Contact with government research organisations
(9) Consultancy arrangements with (7), (8) and independents
(10) Acquisition of other firms or mergers
(11) Joint ventures
(12) Co-operative research arrangements
(13) Licensing and cross-licensing of new products and processes and know-how transfer agreements
(14) Contract research
(15) Other

or some combination of (11) to (14) was often necessary to realise the potential benefits of these ideas. Especially in the improvement of processes, collaboration with the plant contracting and construction industry was also important. Some chemical firms had their own contractor subsidiaries (e.g. Hoechst and Uhde) but independent contractors, such as Scientific Design and UOP, were also important (Freeman *et al.*, 1968). Most contractors, however, were much weaker in R&D than the chemical firms themselves.

The most successful firms in the world chemical industry were characterised by a capacity *both* to make major *radical* innovations and to follow through with numerous *incremental* process innovations and applications innovations. These were often associated with scaling up of plant, which was sometimes done jointly with plant contractors or with licensees. In the case of product innovations they were often the result of joint activity with users in enlarging the range of applications by specific modifications to the product. In both cases this enhanced the development of strong firm-specific technological competence, extending from R&D to design-engineering, production control and marketing to various types of technical service. By the 1960s the strongest firms, such as BASF, employed a very high proportion of technically qualified people and only a minority were actually engaged in production.

## 2. Empirical research on innovation success

Until the 1960s most studies of innovation were mainly anecdotal, based on the personal recollections of leading engineers, managers or scientists. Although economists had always recognised the great importance of innovations for productivity growth and for the competitive performance of firms, they did not make any empirical studies of innovative activities or even of the diffusion of innovations. Even those economists, such as Schumpeter, who put innovation at the centre of the entire theory of economic growth and development did not study the specific features of actual innovations in any depth. He attributed innovative success to a general quality of 'entrepreneurship' but recognised that with the growth of large monopolistic firms the nature of this activity had radically changed (Schumpeter, 1928, 1939).

Nevertheless, we owe to Schumpeter the crucial distinction between *inventions* and *innovation*, and the recognition that innovation is the vital ingredient of dynamic competition. However important in themselves, *inventions* do not have any economic effects unless they are developed to the point of commercial viability and launched on the market. Schumpeter was undoubtedly right to insist that an act of entrepreneurship is essential to translate inventions into successful new products and processes and to recognise the creative aspects of this activity. Although he did recognise that entrepreneurship could be a collective, or even a state activity, he put the main emphasis on the heroic individual entrepreneur. And although he identified the growth of professional in-house R&D as a fundamental change in the organisation of large-scale industry, he did not examine the interaction between the R&D function and other established functions within the firm. Thus it remained as a task for his successors to put flesh and blood on the bare bones of his concept of entrepreneurship.

It was not until the 1960s that a more systematic, empirical approach to innovation began, and at about the same time empirical surveys of diffusion of innovations also took off. In this period in the 1960s and early 1970s most, if not all, of the work was concentrated on the study of specific individual innovations. It aimed to identify those characteristics of each innovation which led to commercial as well as to technical success, whilst recognising that an element of technical and commercial uncertainty is inherent in this activity.

The most effective way to identify those factors which are important for success is to compare those innovations which succeed with those which fail. This was the main feature of one of the most comprehensive empirical studies of innovations: Project SAPPHO (Rothwell, 1985; Rothwell *et al.*, 1974). This project measured about a hundred characteristics of about forty pairs of innovations — forty successes and forty failures. Only about a dozen of the many possible hypotheses systematically discriminated between success and failure. The most important of these were the following (Table 2).

**Table 2** SAPPHO Project: profile of successful innovations

(1) Understanding of user needs
(2) Integration of R&D, production and marketing functions
(3) Linkages with external science and technology network
(4) Concentrated high-quality R&D
(5) Powerful, experienced, senior business innovator
(6) Basic research within enterprise (chemicals only)

*Source:* Freeman (1982); Science Policy Research Unit (1972)

### *User needs*

Successful innovators were characterised by determined attempts to develop an understanding of the special needs and circumstances of potential future users of the new process or product. Failures were characterised by neglect or ignorance of these needs.

### *Coupling of development, production and marketing activities*

Successful innovators developed techniques to integrate these activities at an early stage of the development work. Failures were characterised by the lack of adequate internal communications within the innovating organisation and lack of integration of these functions.

### *Linkage with external sources of scientific and technical information and advice*

Successful innovators, although typically having their own in-house R&D, also made considerable use of other sources of technology. Failures were characterised by the lack of communication with external technology networks, whether national or international.

*Concentration of high-quality R&D resources on the innovative project*

Whereas size of firm did *not* discriminate between success and failure, size of R&D *project* did discriminate. Moreover, the innovations which failed not only had lower resources than those which succeeded but also suffered from failures in development leading to lower quality products. Both *quantity* and *quality* of R&D work thus complemented external networks.

*High-status, wide experience and seniority of the 'business innovator'*

The term 'business innovator' was used in the project to describe that individual who was mainly responsible within the firm for the organisation and management of the innovative attempt. He thus corresponded most closely to the Schumpeterian 'entrepreneur'. Contrary to the original expectations of the Sappho researchers, this individual was generally older in the innovations which succeeded than in those which failed. This result was interpreted as indicating that, particularly in large organisations, innovation could not succeed without the strong commitment of top management to the project and that the role of co-ordination was very important.

*Basic research*

The performance of in-house oriented basic research was associated with success, particularly in the chemical industry.

The original SAPPHO Project was concentrated on only two branches of manufacturing industry — chemicals and scientific instruments. But later research in several other countries not only confirmed the main results but showed that they also applied to other sectors such as machinery and electronics (e.g. Rothwell, 1977; Maidique and Zirger, 1985; Mueser, 1985; Teece, 1987).

However, as the SAPPHO researchers were themselves at pains to emphasise, their method had certain limitations. In particular it did not take into account some characteristics of the *firm* or the *country* in which the innovation attempt took place. It concentrated on single innovations and did not attempt to explain why some firms had a number of successes over an extended period; or why innovations were relatively more successful in certain *countries* over long periods. The SAPPHO Project was international in the sense that it included innovative attempts in many European countries and the United States, but it did not include Japan, nor did it attempt to examine inter-country differences in performance. Consequently, subsequent research has concentrated increasingly on *inter-firm* differences in performance and also on *inter-country* differences.

## 3. Inter-firm differences in innovative success in the chemical industry

Already in the original SAPPHO Project, some interesting differences had emerged between the *chemical* industry and others. Some of these differences related, of course, not just to innovation but to general characteristics of the industry. The chemical industry is relatively concentrated and dominated in most countries by a few large firms. It is both capital-intensive and R&D-intensive. At the time when the SAPPHO Project was carried out, *process* innovations generally required expensive pilot plant work as well as laboratory research. This sometimes involved collaboration with design-engineering and construction firms. Indeed, some of the innovations were actually made by plant contractors, such as Scientific Design or DSM. More commonly, however, they were made by large chemical firms, such as BASF or ICI, who had the ability to design and engineer large plants, as well as to make product innovations. In general the most important external sources of scientific and technical information, advice and collaboration were universities and design-engineering contractors (i.e. [4] and [7] in Table 1).

Some of the innovations attempted by such firms were, of course, failures. There is no way in which success can be guaranteed. An element of risk is always present with innovation. Thus, for example, the SAPPHO Project studied DuPont's failure with CORFAM, estimated to have cost $500 million, and ICI's similar failure with ORTIX, a synthetic leather material, which, however, did not go to the stage of full commercial production and was far less expensive.

Nevertheless, despite occasional failures such as these, it is notable that over a long period some chemical firms, including these two and the three German chemical firms originally linked in the IG Farben Trust, have had a remarkable series of successful innovations in new products and processes, which have enabled them to remain in the forefront of the world chemical industry. It is therefore clearly of great interest to identify features of innovation management in these firms which may have affected their long-run performance.

The first SAPPHO Project was carried out over several years (1968–71) by a mixed team of scientists, engineers and economists. Fortunately, it included a chemist, Basil Achilladelis, who, after leading the SAPPHO work on chemical processes, was able to continue his work on innovation in the chemical industry in the United States in the 1980s. He was exceptionally well qualified for this research, as he had worked as a production manager in the chemical industry in addition to obtaining two doctoral qualifications — one in chemistry and one in economics of technical innovation (Achilladelis, 1973).

Together with other colleagues at the University of Oklahoma, with the support of the National Science Foundation, Achilladelis has completed a project on 'Innovation and the Firm in the Chemical Industry — which included an interesting and thorough study of innovation in the worldwide pesticides industry over a very long period — from the 1930s to the 1980s (Achilladelis *et al.*, 1987). The results of this research enable us to extend the analysis of successful chemical innovations beyond the level of individual projects studied in SAPPHO to the level of the chemical firm (Achilladelis *et al.*, 1990).

Already in the SAPPHO Project it had been observed that the performance of *basic research* within the enterprise was associated with innovation success (Table 2). The association was not a strong one except in the chemical industry where it was important. Furthermore, it had also been observed that success was

**Table 3** Concentration and technological accumulation in chemical innovations 1930–80*

| (*1*) | (*2*) | (*3*) | (*4*) |
|---|---|---|---|
| *Type of innovations and companies* | *Pesticides* | *Pesticides* | *Synthetic materials* |
| | *Top 5 companies* | *Top 10 companies* | *Top 5 companies* |
| | Bayer<br>Geigy<br>ICI<br>Dow<br>DuPont | Col. (2) plus:<br>BASF<br>Hoechst<br>Shell<br>Cyanamid<br>Sumitomo | Bayer<br>BASF<br>Hoechst<br>DuPont<br>ICI |
| Per cent of all innovating companies | 6 | 12 | 5 |
| Per cent of all product and process patents | 19 | 27 | 30 |
| Per cent of all new products | 31 | 44 | 58 |
| Per cent of all radical innovations | 38 | 54 | 60 |
| Per cent of major market successes | 35 | 55 | 66 |

* Synthetic materials 1930–55
*Sources:* Achilladelis *et al* (1987) Freeman (1963, 1982).

associated with the capacity to find more radical solutions to scientific and engineering problems. Patenting activity, on the other hand, did not appear to be associated with success. Both successful and unsuccessful attempts to innovate were associated with patenting, and a greater number of patents did not appear to be a characteristic of success.

In their study of the pesticide industry (which included fungicides and herbicides as well as insecticides), Achilladelis and his colleagues found that only a few companies dominated worldwide radical innovation over a very long period — ten companies accounted for 44 per cent of 846 new products from 1930 to 1980, and for over 50 per cent of those which were the most radical innovations and the most successful on the market (Table 3).

An earlier study (Freeman *et al.*, 1963) had shown a similar concentration of innovation success in the synthetic materials industry from 1930 to 1955. Both this study and the Achilladelis study thus demonstrated clearly the dominant role of a few leading companies in the world chemical industry over long periods. Both studies attempted to explain this prolonged period of innovative success and pointed to the central importance of '*technological accumulation*' — the development of a strong research and innovation tradition within the firm in key specialised areas. Such accumulated strength and experience appears to outlive particular individuals and business innovators, although the outstanding qualities of individual managers are, of course, also important. Success breeds success because the skills and competence generated in research, production and marketing in one major breakthrough radical innovation breed confidence and prestige and facilitate further innovations on the same technological trajectory. Many of these will be incremental applications innovations but some will be further radical innovations based on similar scientific foundations. Pavitt (1984, 1986) has shown that this process of firm-specific technological accumulation is characteristic of science-based industry.

The German chemical firms and some Swiss chemical firms had a ratio of R&D expenditure to sales of over 5 per cent and sometimes higher than 7 per cent for most of the 1920s and 1930s. This was substantially higher than most of their (less successful) competitors. In the post-war period the leading firms have continued to maintain a high ratio of R&D/sales, but it was only in the 1980s that major Japanese firms reached this level (between 4 and 5 per cent).

The leading firms have a share of radical innovations which is much higher than their share of patents. This can be explained in terms of the need of all the imitating and overtaking firms to secure 'me-too' patents. Achilladelis (1987) found that large numbers of patents did not necessarily lead to radical innovations in pesticides. The earlier study of synthetic materials showed that the few leading firms had a significantly larger share of both *key* patents and innovations than of patents in general. But the firms which make radical innovations all had significant numbers of patents.

Achilladelis pointed out that the top five companies in the world pesticide industry had already made their first radical innovations in the 1930s or 1940s

**Table 4** Some examples of corporate technological tradition and early innovation

| | | | |
|---|---|---|---|
| *Pesticides* | Dow | Pentachlorophenol (Pesticides) | 1930 |
| | DuPont | Nabam (Fungicides) | 1936 |
| | Ciba-Geigy | DDT (Insecticides) | 1939 |
| | ICI | MCPA (Herbicides) | 1942 |
| | Bayer | Parathion | 1942 |
| | Monsanto | Randox (Herbicides) | 1955 |
| *Other products* | Bayer | First Synthetic Rubber | 1910 |
| | Bayer | Polyurethanes | 1942 |
| | BASF | Ammonia Synthesis | 1912 |
| | BASF | Polystyrene | 1928 |
| | Dow | Polystyrene | 1932 |
| | ICI | Polyethylene | 1936 |
| | DuPont | Nylon | 1936 |

*Source:* Achilladelis *et al.* (1990)

(Table 4). These early successes were sometimes related to even earlier or parallel innovations in organic chemical intermediates and dyestuffs. All the leading companies in the innovation of synthetic materials in the 1930s were also among the leaders in pesticide innovation. Strength in organic chemistry research also facilitated other product and process innovations.

It is easy to see the connections between corporate technological traditions and the factors identified in Project SAPPHO as being associated with innovation success. The experience of one or two successful innovations enhances understanding of the market, even if new users are involved. The linkages with external sources of scientific and technical expertise are greatly strengthened by successful innovation, both through recruitment of young scientists and engineers and through recognition in the world scientific and technical community. Thus, high-quality R&D is further strengthened and reinforced within the company.

However, there are times when the very strength of a particular technological tradition may inhibit further innovation. This occurs when a trajectory is approaching exhaustion and reaching the limits of its profitable exploitation. In these circumstances the methods, attitudes and traditions engendered by past success may actually prove a barrier to the adoption of new approaches and new trajectories. Such fundamental changes in technology are sometimes described as changes in 'technological paradigm' following the analogy with Kuhn's (1961) analysis of paradigm change in science. On the basis of this analogy, incremental innovation may be viewed in the same way as Kuhn's 'normal' science: a cumulative process of knowledge generation on fairly well-established principles. Paradigm change involves new departures in ideas, in attitudes and in organisation, whether in science or in technology.

From the 1930s to the 1970s one of the strongest technological trajectories was based on macro-molecular chemistry, following Staudinger's work on long-chain molecules in the 1920s. This led to the introduction of entire families of new materials, including fibres and rubbers. These products (together with drugs, fertilisers and pesticides) were the fastest growing sectors of the chemical industry from the 1930s. Numerous product and process innovations were made throughout the 1930s and 1940s and continued to the 1950s (Freeman, 1963). However, the pattern of innovation in the 1960s shifted increasingly from radical product to process innovations (Walsh, 1984). There were also other indications in the 1960s that this trajectory was approaching its limits.

The recent MIT study of US industry (Dertouzos, Lester and Solow [eds], 1989) sums it up as follows:

> Technological maturity. By the 1970s the chemical industry had matured technologically and had become largely a commodity business. Process technology was roughly comparable worldwide. New entrants into the market, particularly the large oil companies, led to sharpened competition and lower profit margins. This discouraged large R&D expenditures with distant and uncertain payoffs. Feedstocks were cheap (until the 1973 oil shortage), and there were large, complex plants in place to process them. These factors also shifted incentives away from significant new R&D efforts towards maximizing current production and return on existing investment. Some

companies also diversified outside the chemical business. All of these moves reduced innovation in the industry. [p. 193]

Perhaps the clearest indication that the old paradigm was reaching its limits in the chemical industry came from the empirical work of Baily and Chakrabarti (1988). They found that the average number of chemical *product* innovations in the US chemical industry fell dramatically in the 1970s. Process innovations also declined, but much less steeply (Table 5). These results confirm the analysis made in the MIT report.

**Table 5** Innovation in the US chemical industry 1967–82 (average number per annum)

| *Period* | *Products* | *Processes* | *Equipment* |
|---|---|---|---|
| 1967–73 | 332 | 39 | 108 |
| 1974–79 | 39 | 32 | 57 |
| 1980–82 | 65 | 35 | 105 |

*Source:* Baily and Chakrabarti (1988)

The decline in the number of innovations in the US chemical industry was associated with a more widespread change in technology, affecting many other industries. Such breaks with past technological trajectories, which are economy-wide in their impact, have been described as changes in 'techno-economic paradigm' (Perez, 1983). They involve the rise of new 'best practice' for management and many organisational as well as technical innovations. They also create opportunities for the rise of new firms, industries and countries to technological leadership.

The next section discusses the characteristics of the new techno-economic paradigm, which is displacing the old oil-intensive mass and flow production systems characteristic of US industry from the 1930s to the 1970s.

## 4. Change of techno-economic paradigm and new patterns of innovation

A number of economists have pointed to the importance of 'technological trajectories' (Nelson and Winter, 1977) and of 'constellations of innovations' (Keirstead, 1948), which are both technically and economically interrelated. Several have also extended Kuhn's (1961) notion of scientific paradigms to the concept of 'technological paradigms' (Dosi, 1982). Nelson and Winter (1977) also suggested that some trajectories could be so powerful and influential that they could be regarded as 'generalised natural trajectories'. They suggested electricity as one such example. However, Carlota Perez (1983) was the first to take these scattered ideas and comments and to relate them not just to a particular branch of industry but to the broad tendencies in the economy as a whole. Her theory may be described as a 'meta-paradigm' or a 'pervasive technology' theory. In this way she gave some real content to the notion of 'successive industrial revolutions'.

A change of techno-economic paradigm or 'technological style' brings with it a whole range of new products and processes and many others which are redesigned to take advantage of the new technical and economic possibilities. Perez suggests that underlying this paradigm change is not just a range of new products but a change in the dynamics of the relative cost structure of all possible inputs into production. In each new techno-economic paradigm a particular input or set of inputs, which may be described as the 'key factor' or factors of that paradigm, fulfil the following conditions:

(i) *Clearly perceived low and rapidly falling relative cost.* As Rosenberg (1976) and other economists have pointed out, *small* changes in the relative input cost structure have little or no effect on the behaviour of engineers, designers and researchers. Only major and persistent changes have the power to transform the decision rules and 'common sense' procedures for engineers and managers (Perez, 1985; Freeman and Soete, 1987). The oil 'shocks' of 1973 and 1979 are examples of such changes. So, too, are the reductions in the cost of integrated circuits in the 1970s and 1980s.

(ii) *Apparently almost unlimited availability of supply over long periods.* Temporary shortages may, of course, occur in a period of rapid build-up in demand for the new key factor, but the prospect must be clear that there are no major barriers to an enormous long-term increase in supply. This is an essential condition for the confidence to take major investment decisions which depend on this long-term availability.

(iii) *Clear potential for the use or incorporation of the new key factor or factors in many products and processes throughout the economic system*; either directly or (more commonly) through a set of related innovations, which both reduce the cost and change the quality of capital equipment, labour inputs, and other inputs to the system.

Perez maintains that this combination of characteristics holds today for microelectronics and few would deny this. It held until recently for oil, which underlay the post-war boom (the 'Fourth Kondratieff' upswing). Before that, she suggests that the role of key factor was played by low-cost steel in the Third Kondratieff (1880s to 1930s).

Clearly, every one of these inputs identified as 'key factors' existed (and was in use) long before the new paradigm developed. However, its full potential was only recognised and made capable of fulfilling the above conditions when the previous paradigm and its related constellation of technologies gave strong signals of diminishing returns and of approaching limits to its potential for further increasing productivity or for new profitable investment. As we have seen, this was the case with oil-based energy-intensive technologies in the 1970s.

The new key factor does not appear as an isolated input, but rather at the core of a rapidly growing system of technical, social and managerial innovations, some related to the production of the key factor itself and others to its utilisation. Gradually the new model of innovation affects all sectors of the economy. It embodies three main features (Perez, 1985) in the case of the new information technology paradigms:

(*1*) Information-intensity rather than energy-intensity in products and processes. In the case of Japan this shift has been studied and confirmed by a number of researchers.

(*2*) Flexibility in process investment and in product mix rather than the dedicated capital equipment and standardised products range characteristic of the 'Fordist' mass-production system.

(*3*) 'Systemation' not just 'automation'. Fordist production systems already used automated production equipment and process plant. The new paradigm uses information technology and communication technology to integrate the production plant with the design, marketing and administration functions within the firms.

At first sight it might appear that this change of paradigm, although certainly heavily affecting the electronics and machinery industries, has no direct relevance to the chemical industry. But this would be a very superficial impression. The new paradigm affects the chemical industry profoundly in many ways. Among the more important are the following.

First of all, already for some time the process instrumentation of the industry has been deeply affected by the development of new electronic instruments. Successive generations of these instruments based on new generations of integrated circuits have permitted many energy and material-saving innovations and greater precision in process control generally. They have also facilitated a shift away from the gigantic dedicated plant to more flexible multi-product plants responding more quickly to market changes and the cost of inputs. The Baily and Chakrabarti study showed that *instrument* innovations had sharply *increased* in the US chemical industry in the 1980s.

Secondly, chemical firms, like all others, are deeply affected by the capacity of computer technology to integrate various functions within the firm and by the capacity of new communication technology to link together diverse production sites, R&D laboratories and other activities within the firms. Chemical engineering activities and especially the design function become more flexible and again permit more rapid changes in both products and processes. (See also Table 1 in Chapter 6.)

Thirdly, the fastest growing sectors of demand, and especially the electronics industry itself, require more speciality high-grade products rather than bulk commodity materials. The new materials technology, based on the intensive use of information technology in research and in computer-aided design (CAD), permits a very wide range of choice in materials, tailored to most customer-specific applications. Special plastics, such as polymer alloys and polyimides, and composite materials will increase their share of production at the expense of the bulk commodity materials (Kasama, 1989). They will increasingly be sold as a package, with services to meet customers' specialised requirements. (See also the chapter by Cobbenhagen, den Hertog and Philips in this volume.)

Thus the change in techno-economic paradigm does already deeply affect the chemical industry and will do so even more in the 1990s. The final section of this chapter discusses the way in which the *national* system of innovation may affect the capacity of industry to adapt to the change of paradigm.

## 5. Conclusions: change of paradigm and national systems of innovation

Changes of techno-economic paradigms involve major institutional changes in society. The management systems at the level of the firm, the industry and the nation, which have been developed to promote one type of technology, are generally inappropriate for entirely new and different technologies. In fact, one of the main problems confronting the older industrial countries in changing techno-economic paradigm is precisely the inertia of some established institutions. This is made abundantly clear in the MIT study on the problems of US manufacturing industry (Dertouzos, Lester and Solow [eds], 1989). Similar problems arose for the British economy at the end of the last century as the leading countries at that time shifted to a new paradigm based on electricity and cheap steel.

In the shift to a new techno-economic paradigm based on information and communication technology the Japanese economy has some special advantages and has outstripped the USA and the older industrial countries of Europe in many areas of application, such as robotics and consumer electronics. In fact, in the key sector of computers, electronic components and telecommunications, the Japanese electronics industry has become the world leader in international trade. To a considerable extent this shift in technological and economic leadership from the USA to Japan was based on institutional innovations in Japanese industry and government (Freeman, 1987).

Particularly important has been the technique of managing product and process innovations. Japanese work on innovation management stresses very strongly the integration of R&D, production and marketing and various social techniques for achieving this. Takeuchi and Nonaka (1986) described what is needed as 'Stop running the relay race and take up rugby.' Aoki (1986) describes it as 'horizontal information structures'. Japanese firms have been particularly adept at using such techniques to increase the flexibility in their product mix and improve the speed and quality of their new product development. This type of social innovation is closely related to Japanese methods of recruitment and training which facilitate the integration of functions within the firm. Finally, Japanese government agencies have been alert to the importance of information and communication technology for quite a long time and have adopted many policies to improve the infrastructure and stimulate an adequate response to the change of paradigm. For historical reasons the Japanese economy was never so committed to Fordist systems as the USA and European countries were.

In the chemical industry the performance of the Japanese economy has not been quite so strong as in electronics. Following the work of the MITI Petrochemical Technology Committee in 1954, the petrochemical industry developed rapidly on the basis of licensed technology and in some cases joint ventures. By 1970 Japanese firms were already making many independent improvements in the design of the processes which they had imported and had been able to move very rapidly from importing a process to exporting an improved process (Saffer and Yoshida, 1980). The growing strength of Japanese incremental innovation in the petrochemical industry and in many other areas of chemical technology is clearly shown by the huge increase in Japanese patenting in the 1970s.

However, partly because of dependence on imported oil and the cost advantage of some petrochemical processes in the Middle East, Japan developed a deficit in foreign trade in chemicals. Moreover, some observers have maintained that Japanese chemical firms are still behind the leading European and US chemical firms in scale of R&D and in radical innovations, despite the very strong performance in patenting, e.g. Achilladelis *et al.* (1990). According to this view the strength of basic research in Europe and the USA will be an additional factor favouring the European and US chemical industries. As we have seen, basic research has been important for the chemical industry for over a century already. The emergence of biotechnology as a new fundamental source of innovations is highly relevant here and has renewed the importance of university–industry links (Faulkner, 1985). Some observers have suggested that this area of university–industry linkage is a particularly strong feature of the American national system of innovation (Nelson, 1988). This could possibly be a source of relative Japanese weakness during the paradigm change since many radical product and process innovations are expected to emerge in the 1990s on the basis of biotechnology.

As against this view it can be argued that the Japanese chemical industry is now well placed to overtake the European and US firms for the following reasons:

(*1*) The change of techno-economic paradigm is leading to a major change in the materials required in the economies of the industrialised countries. As we have seen, this means a shift from commodity chemicals to speciality 'customised' chemicals and a varying product mix from flexible plants. Much of the demand for new materials comes from the electronics industry and from those sectors of the machinery and vehicles industry deeply affected by microelectronic technology.

The enormous strength of these industries in Japan means an assured market and a strong user–producer interface with very high quality standards. Whilst 'demand-pull' is certainly not sufficient in itself to ensure innovative success (Wiseman, 1983; Walsh, 1984) it is a very important factor interacting with research strength and technology push. A particularly important feature of Japanese technological strength over the past twenty years has been the phenomenon of 'fusion' in industrial R&D (Kodama, 1986) as in 'mechatronics'. The development of 'chematronics' is likely to be just as important, and even though this may mean some new entrants (just as oil firms entered the industry in the previous period), it will vastly strengthen the Japanese chemical industry. Similar developments are, of course, occurring in Europe and the United States, but they are both more rapid and more widespread in Japan. According to MITI estimates (Kasama, 1989), demand for these new materials by the year 2000 may amount to the equivalent of a quarter of the entire present output of the Japanese chemical industry. The strong initiatives of MITI in ceramics research and other new materials have also further stimulated this shift.

(*2*) It is, of course, true that biotechnology will also be a very important source of new materials and processes. But it seems likely that the main areas of application will be in agriculture and in health services. The early hopes of substitution for established industrial chemical processes and for such bulk commodities as animal feeding stuffs have not been fulfilled and are unlikely to be realised unless there is a drastic change in cost structure of inputs (OECD, 1988). Nevertheless, this is certainly an area where radical new possibilities are opening up and the unexpected can happen. Moreover, the drug industry and agricultural products are two of the fastest growing sectors of demand. Most of the leading European and US chemical and drug companies, such as ICI, Bayer, Monsanto and Pfizer, have certainly long since recognised the extraordinary long-term potential of biotechnology and have reoriented the work of their R&D departments accordingly, as well as developing new links with external sources. Some breakthrough innovations have been made by smaller new biotech firms in Europe and the USA. Again, the capacity to launch innovative entrepreneurial small firms in high-tech industries is often claimed as one of the strengths of US industry. However, it would be a great mistake to belittle Japanese capacity in biotech research. The OECD study showed that many more firms in a variety of industries (including chemicals, of course) had launched biotech R&D programmes in Japan than in any other country. Japan already has a strong research tradition in fermentation technology. Finally, Japanese firms have demonstrated the capacity rapidly to take up new developments in any branch of industry even if they are launched initially in other countries. Pavitt (1986) has argued that this capacity of large firms to adapt to the potential of new technologies has made Schumpeter's idea of creative *destruction* obsolete.

(*3*) Whilst it is true that strong links with fundamental research are extremely important for innovation in the chemical industry, a large part of this research is in the public domain and open to researchers all over the world. Japanese contributions to this published research are still proportionately much lower than they are to world patents (Irvine and Martin, 1986; Freeman, 1987). Nevertheless, Japanese university researchers and industrial researchers are making an increasing contribution and have already demonstrated their capacity to make innovations based on this research for many decades. It is not actually necessary to be the *leader* in basic research to be leader in technology, as the USA demonstrated between the 1880s and 1930s. Finally, Japanese chemical firms as well as Japanese electronics firms have been strengthening their in-house basic research facilities in the 1980s much more rapidly than their European or US competitors.

(*4*) As we have seen, adaptation to a new pattern of technological development depends on organisational, managerial and institutional changes. The Japanese style of innovation management seems particularly well adapted to the characteristics of the new techno-economic paradigm, both with respect to the integration of various functions within the firm and with respect to 'fusion research' (Kodama, 1986). As the MIT study of American industry stresses, co-operation with other firms and with government is more characteristic of Japanese firms than of US firms. The ability of large Japanese conglomerates to adapt new technology and other general characteristics of the Japanese environment will benefit every industry in Japan, including the chemical industry.

It is, of course, not possible to make any accurate predictions about such complex social problems far into the future and the above comments are necessarily speculative. However, they do raise this interesting question: does the rise of a new national system of innovation affect all industries and services in a change of paradigm, or does it affect only a few leading industries? The experience of the world chemical industry in the 1990s and the early twenty-first century will be a fascinating test of the proposition that national circumstances generally predominate over the particular firm or industry in determining innovative success. The European and US firms are certainly not standing still. They are changing rapidly (see, for example, Marsh, 1989 on ICI) and are learning from their own past successes and failures. If they are nevertheless overtaken by Japanese chemical firms over the next twenty years, this would be through a combination of the efforts of the Japanese firms themselves and the national system of innovation in Japan.

## References


Achilladelis, B.G. (1973), 'Process Innovation in the Chemical Industry', D. Phil. Thesis, University of Sussex.

Achilladelis, B.G. *et al.* (1987), 'A study of innovation in the pesticide industry', *Research Policy*, 16 (2), pp. 175–212.

Achilladelis, B.G. *et al.* (1990), 'The dynamics of technological innovation: the case of the chemical industry', *Research Policy*, 19 (1), pp. 1–35.

Aoki, M. (1986), 'Horizontal vs vertical information structure of the firm', *American Economic Review*, 76 (5), pp. 971–83.

Baily, M.N. and A.K. Chakrabarti (1988), *Innovation and Productivity in US Industry*, Brookings Institution.

Beer, J.J. (1959), *The Emergence of the German Dye Industry*, University of Illinois Press.

Beer, J.J. (1966), 'Why look back?' the function of history in industrial research organisations', *Research Management*, 9 (2), pp. 101–7.

Bradbury, F.R., McCarthy, M.C. and C.W. Suckling (1972), 'Patterns of innovation', *Chemistry and Industry*, 316, pp. 22–6.

Dertouzos, M., Lester, R. and R. Solow (eds) (1989), *Made in America: Regaining the Productive Edge*, MIT Press.

Dosi, G. (1982), 'Technological paradigms and technological trajectories', *Research Policy*, 11 (3), pp. 147–62.

Dosi, G., Freeman, C., Nelson, R., Silverberg, G. and L. Soete (eds) (1988), *Technical Change and Economic Theory*, New York: Columbia University Press/London: Pinter.

Faulkner, W. (1985), 'Linkage between industrial and academic research: the case of bio-technology research in the pharmaceutical industry', D. Phil. Thesis, University of Sussex.

Freeman, C. (1982), *The Economics of Industrial Innovation*, London: Pinter.

Freeman, C. (1987), *Technology Policy and Economic Performance: Lessons from Japan*, London: Pinter.

Freeman, C. *et al.* (1963), 'The plastics industry: a comparative study of research and innovation', *National Institute Economic Review*, 26, pp. 22–60.

Freeman, C. *et al.* (1968), 'Chemical process plant innovation and the world market', *National Institute Economic Review*, 45, pp. 29–57.

Freeman, C. and L. Soete (eds) (1987), *Technical Change and Full Employment*, Oxford: Blackwell.

Gibbons, N. and R. Johnston (1974), 'The roles of science in technological innovation', *Research Policy*, 3 (3), pp. 220–42.

Irvine, J.H. and B. Martin (1986), 'Is Britain spending enough on science?' *Nature*, 323, pp. 591–594.

Jewkes, J., Sawers, D. and J. Stillerman (1958), *The Sources of Invention*, London: Macmillan.

Kasama, Y. (1989), 'Japan warned of danger of relying on commodities', *European Chemical News*, 3 April, pp. 26–7.

Kelly and Kranzberg (eds) (1978), *Technological Innovation: a Critical Review of Current Knowledge*, San Francisco Press.

Keirstead, B.G. (1948), *The Theory of Economic Change*, Toronto: Macmillan.

Kodama, F. (1986), 'Japanese innovation in mechatronics technology', *Science and Public Policy*, 13 (1), pp. 44–51.

Kuhn, T. (1961), *The Structure of Scientific Revolutions*, Chicago University Press.

Maidique, M.A. and B.J. Zirger (1985), 'The new product learning cycle', *Research Policy*, 14, pp. 299–313.

Marsh, P. (1989), 'Why ICI believes people need space', *Financial Times*, 9 August, p. 12.

Mueser, R. (1985), 'Identifying technical innovations', *IEEE Transactions on Engineering Management*, EM-32 (4), pp. 158–76.

National Science Foundation (1973), *Interactions of Science and Technology in the Innovation Process*, NSF 667, Washington.

Nelson, R. (1988), Chapter 15, 'Institutions supporting technical change in the United States' (eds) Dosi *et al.*, *Technical Change and Economic Theory*, London: Pinter.

Nelson, R. and S.G. Winter (1977), 'In search of a useful theory of innovation', *Research Policy*, 6, pp. 37–76.

OECD (1988), *Bio-Technology*, (Wald Report), Paris: OECD.

Pavitt, K. (1984), 'Sectional patterns of technical change: towards a taxonomy and a theory, *Research Policy*, 13 (6), pp. 343–73.

Pavitt, K. (1986), 'Chips and trajectories', in MacLeod, R. (ed.), *Technology and the Human Prospect*, London: Pinter.

Perez, C. (1983), 'Structural change and the assimilation of new technologies in the economic and social system', *Futures*, 15 (4), pp. 357–75.

Perez, C. (1985), 'Micro-electronics, long waves and world structural change', *World Development*, 13 (3), pp. 441–63.

Rosenberg, N. (1976), *Technology in Perspective*, Cambridge University Press.

Rothwell, R. (1977), 'The characteristics of successful innovation and technologically progressive firms', *R&D Management*, 7 (3), pp. 191–200.

Rothwell, R. (1985), 'Project SAPPHO: a comparative study of success and failure in industrial innovation', *Information Age*, 7 (4), pp. 215–19.

Rothwell, R. *et al.* (1974), 'Sappho updated', *Research Policy*, 3 (3), pp. 257–91.

Saffer, A. and J. Yoshida, (1980), 'Sources of technology', *Chemtech*, pp. 670–3.

Schumpeter, J.A. (1928), 'The instability of capitalism', *Economic Journal*, Vol 38, pp. 361–86.

Schumpeter, J.A. (1939), *Business Cycles*, 2 vols, New York: McGraw Hill.

Science Policy Research Unit (1972), *Success and Failure in Industrial Innovation*, London: Centre for the Study of Industrial Innovation.

Takeuchi, H. and I. Nonaka (1986), 'The new product development game', *Harvard Business Review*, January–February, pp. 137–45.

Teece, D.J. (ed.) (1987), *The Competitive Challenge: Strategies for Industrial Innovation and Renewal*, New York: Ballinger.

van de Ven, A.H. (1988), 'Progress Report on the Minnesota Innovation Research Program', University of Minnesota, Hubert Humphrey Institute of Public Affairs.

Walsh, V. (1984), 'Invention and innovation in the chemical industry: demand-pull or discovery-push?', *Research Policy*, 13 (4), pp. 211–34.

Wiseman, P. (1983), 'Patenting and invention activity in synthetic fibre intermediates', *Research Policy*, 12 (6), pp. 329–41.

# Part II: Diffusion of innovations and productivity growth

# 5. The character of technological change and employment in banking: a case-study of the Dutch automated clearing house (BGC)

*G.R. de Wit*

## 1. Introduction

Two fields of economic analysis have been neglected in the past: the study of *technological change*; and the analysis of the *services sector*. However, the economic significance of the services sector in terms of employment is great and the overwhelming part of computer and communication items of the capital stock are located in this branch as well. The analysis of the impact of technological change on employment in the services sector, however, is lacking.

Banking is probably the first major services branch which has adopted new information technologies extensively. In this chapter, we will address the character of technological change and employment in banking. Before analysing the process of technological change in more detail, we will first address the 'stylised facts' of the services sector, the implementation and use of information technologies and its labour characteristics in the next section. Subsequently, Section 3 outlines traditional models of technological change as well as an adapted model in order to analyse the process of technological change and employment in a services branch such as banking. Section 4 applies the model to the case of the Dutch automated clearing house as a representative firm of mainframe computer technologies in banking. After an analysis of the *rate* of technological change at the BGC (the automated clearing house) a more detailed description of the *nature* (not just the conventional classification of labour-saving, neutral or capital-saving) will be presented as well. The chapter is summarised in Section 5.

## 2. The services sector

### 2.1 'Stylised Facts'

During the course of economic development the shift in the structure of employment towards the services sector has been a persistent feature. The characteristics of the shift can be shown by sectoral changes in employment, expenditure, output, prices and productivity. From 1870 onwards, the share of the Dutch services sector in total employment grew from 34 per cent to 48 per cent in 1960 and to 67 per cent in 1984. Value added in services as a percentage of total gross domestic product increased from 49 per cent in 1960 to 63 per cent in 1984 (Elfring, 1988).

Explanations of this secular shift usually address factors such as changes in income per head, changing consumer tastes and the development of productivity growth. Especially in the mid-1960s, according to Elfring (1988, p. 49), there was a growing belief that lagging productivity growth in services was the main reason for its rising employment share. Two of the most prominent analysts were Baumol (1967) and Fuchs (1964, 1965, 1968 and 1969). Baumol's work was largely theoretical, while Fuchs carried out empirical research.

Baumol's model distinguishes between two sectors: i.e. a goods producing sector, on the one hand; and the services sector, on the other. According to Baumol technological change is predominantly located in the goods producing sector. Following the differences in technological change, the services sector is lagging behind in productivity growth and its share in total employment will increase. Furthermore, under the assumption of a similar wages and salaries development the services sector will be affected by a so-called 'cost disease'. Despite increasing relative price levels in the services sector, Fuchs showed that the higher income elasticities of demand for services compared with manufactured goods resulted in increasing output shares of the services sector.

These 'stylised facts' of the services sector are closely related with traditionally assumed economic characteristics of the branch. Services were assumed to be 'immaterial', low capital intensive and not suited for standardised mass production. Furthermore, services were not homogeneous and could be neither kept in stock nor transported. In contrast, manufactured goods were tangible, could be produced on a large scale, kept in stock and transported over long distances. In other words, the traditional characteristics of services compared with manufactured goods were supposed to explain a great deal of the lagging productivity growth.

The relatively high price level of certain services associated with the unsusceptibility for technological change and low opportunities for productivity growth made Baumol (1967) express his concerns about a possible scarcity of some services in the future. Service industries, however, reacted in many ways to push down the resulting cost pressures. Elfring (1988, p. 67) gives several examples such as the concept of do-it-yourself and the use of information technology. It is the last item which is the focus in this chapter.

### *2.2 Information technology*

According to Elfring (1988, p. 151), the pressure for more efficiency has induced a number of new developments in the services sector. Measures to enhance productivity in commercial services industries today might even be more effective than technological advances in manufacturing. Prospects for continued employment growth in services are increasingly influenced by applications of *information technology*. Many services industries have started to use computer and information technology in order to cure Baumol's 'cost disease'. At the same time, however, some observers view the process of technological change in the services sector as labour-saving and are rather pessimistic about employment growth. This leads Elfring (1988, p. 152) to conclude that some kind of analytical framework is needed in order to assess the employment impact of information technology in services. Such an analytical framework will be developed in the next section. But first, we will sketch the outlines of the implementation of information technologies in the services sector.

Kimbel (1973) expressed his opinion about the *opportunities* of information technologies in the services sector in the early 1970s. He pointed to the vast range of possible applications and identified the government, education, financial services, business services and general information services as five areas very well suited for applications of information technologies. Among them, the *financial services* industry was considered to be most receptive to information technology. As such the branch was one of the first sectors in the economy to introduce applications of computer technology and nowadays it can be considered as a sector already having an outstanding experience in the field of automation and information technologies of around three decades.

Guy examined the growth and changing nature of the *capital stock* in the British services sector. Regarding the structure of the capital stock, it should be noted that the share taken up by plant and machinery in the services sector is far less than in the manufacturing sector. According to Guy (1987, p. 184) in 1983 only 30 per cent of the gross capital stock in the services sector could be defined as consisting of plant and machinery, the remainder consisting primarily of building stock. By contrast, plant and machinery accounted for nearly two-thirds of the manufacturing capital stock. At the same time, however, the relative share of plant and machinery in new investments in the services sector has grown rapidly and was estimated at 50 per cent in 1983. Distinguishing between three broad categories of services the annual growth rates of the plant and machinery capital stock increased from 5.4 per cent in the 1950s to 9.3 per cent in the 1970s and early 1980s in the case of the private commercial services, and decreased from 2.9 per cent to 0.4 per cent in the case of network services and from 13.3 per cent to 5.6 per cent in the case of public social services. The annual growth rate in services as a whole was estimated at 4.0 per cent during the period 1953–83, compared with 3.6 per cent for manufacturing in the same period.

Furthermore, according to Guy, *current investment* in absolute terms in the services sector is nearly three times greater than in the manufacturing sector,

**Table 1** Development of plant and machinery intensities in the British services sector

| | *Plant and machinery capital intensity indices manufacturing 1953 = 100* | | | | |
|---|---|---|---|---|---|
| *Sector* | *1953* | *1963* | *1973* | *1983* | *Growth 1953–83* |
| | | | | | % |
| Network services | 391 | 532 | 908 | 1057 | 3.4 |
| Private services | 19 | 26 | 63 | 140 | 9.2 |
| Public services | 4 | 12 | 25 | 40 | 7.6 |
| Total services | 88 | 104 | 165 | 210 | 2.9 |
| Manufacturing | 100 | 146 | 234 | 413 | 4.9 |

*Source:* Guy (1987) p. 187.

although this is due largely to the extremely high capital intensity of the network services and the role of the financial services sector in the hiring out of plant and machinery to other industrial sectors. An overview of the development of the plant and machinery capital stock per employee in different parts of the British services sector in comparison with manufacturing is given in Table 1.

From Table 1, two main conclusions can be drawn. First, although there are large differences in the plant and machinery intensity between manufacturing and the services sector, investments in plant and machinery, and associated with that processes of technological change, do take place in all parts of the services sector. Second, within the services sector big differences can be identified; the network services, on the one hand, with capital intensities much higher and annual growth rates lower than in manufacturing, and the private and public services, on the other, with lower capital intensities but much higher growth rates. Regarding the relation between technological change and employment the private commercial services are probably the most interesting sub-sector of services in so far as they cover almost one-third of total UK employment and experience the highest growth rate in plant and machinery capital intensity.

Roach (1988) made a comparable analysis for the case of the United States. The average annual growth rate of the capital stock in manufacturing varied between 1.7 per cent in the 1980s and 4.7 per cent in the 1960s. Comparable growth rates for services industries were estimated at 3.6 per cent and 4.5 per cent, respectively, and for the financial industry at 7.8 per cent in the 1980s and 8.7 per cent in the 1960s. Furthermore, he investigated the development of specific investments in computer and communications items. From the total information technology capital stock, around 85 per cent is owned by the services sector and 15 per cent by manufacturing. Within the services sector, it is estimated that around 40 per cent of the whole capital stock is located in the telecommunications sector, 20 per cent in financial services, 10 per cent in trade, 5 per cent in transport and public utilities and the remaining 10 per cent in other private and public services.

### *2.3 Productivity and functionalised technological change*

It will be clear by now that the services sector in general and the commercial services in particular show a high rate of technological change if measured in terms of investments in computer and communications technology items. However, it is far from clear whether the investments in new technology are properly transformed into higher productivity growth rates. In other words, is technological change measured in the classical way of total factor productivity growth increasing? Furthermore, is the character of technological change neutral, labour-saving or capital-saving? In order to assess the impact of technological change on *employment*, it is necessary to gain insight into the *nature* and *direction* as well as the *rate* of technological change as described.

Baily and Chakrabarti (1988) recently addressed a paradox in the field of electronics and white-collar productivity. Paradoxically, rapid electronics innovation as well as a rapid adoption of the new equipment seem to have led to only a minor productivity pay-off, while we would expect a large productivity growth. In order to solve this paradox, they state that the data currently available do not reveal the productivity of information capital (1988, p. 90). Consequently, they plead for better databases. For the present, however, they attempt to explore the paradox through analysis and simulation. They formulate three hypotheses for such simulations. The first is that some output is not measured. Second, that companies engage in activities which are privately, but not socially, productive. Third, that there are incentives affecting the behaviour of individuals which are not in the interest of organisations. Our own approach is a reformulation of the first hypothesis and recognises that investments in information technology and its applications are restricted to specific *functions* within the branch. As a consequence, new investments and the capital stock should be related to a specific part of employment and output only. This will be illustrated by the case of the financial industry.

Gorman (1969) used two different approaches regarding the measurement of productivity in US banking and reached contradictory results. Labour productivity on the basis of the liquidity approach seemed to have been decreasing, while the transactions approach indicated increasing levels of labour productivity. The reason for the contradiction, in our opinion, is the poor relationship between the two different measures of production volume, on the one hand, and the same total labour input used in both approaches, on the other. Because a large part of total employment is associated with the production process of payment transfers, the liquidity approach over-estimates labour input (in relation to the output thus specified) and under-estimates labour productivity. The transaction approach, in contrast, over-estimates production volume growth in relation to labour input (as defined) because the number of transfers grew much faster than, for instance, provisions paid for commercial advice.

Kendrick (1988) estimated total annual factor productivity growth in finance and insurance at 1.1 per cent in the period 1948–73, at –0.7 per cent during the 1970s and at –1.0 per cent in the 1980s. The production volume used by Kendrick was based on valued added figures. This pessimistic outlook on productivity development in banking is confirmed by Roach (1988). He presents figures of –0.1 per cent annual productivity growth in the 1970s and –1.3 per cent in the beginning of the 1980s. Mark (1988) reports about alternative measures of production volume as developed by the Bureau of Labor Statistics in the United States. In the case of the financial industry an alternative was developed along the lines of the transaction approach of Gorman. The production volume was measured as a weighted sum of transactions in the field of deposits, loans and trust services. The partial labour productivity development of the US financial industry reported by Mark was very different from the figures given by Kendrick and Roach. Labour productivity in commercial banking experienced an annual growth of 2.1 per cent during the period 1965–73, 0.2 per cent in the 1970s and 5.4 per cent in the first half of the 1980s. In other words, Mark's figures show a much higher capability of banks to transform investments in information technology items into increases of productivity growth if compared with the results of Kendrick and Roach.

In our opinion, the contradiction results from an insufficient analysis of the history of technological change in the banking industry itself. Although the calculations of productivity growth in the whole banking sector made by Kendrick and Roach are legitimate in themselves, it would be a misconception to interpret the negative productivity rates thus calculated as disappointing results of investments in technology. A misconception would occur because the investments in technology relate neither to the whole production process and production volume of banking nor to the whole input of labour. Nevertheless, implicitly assuming that this would have been the case would imply examining the wrong relationships between the capital inputs (which are associated with a part of the production process only), on the one hand, and the labour input and volume outputs (associated with the whole production process), on the other. Consequently, only productivity growth rates based on analyses of relationships between capital and labour inputs and production volume outputs associated with that part of the production process in which technological change really took place would provide us with the right indicators of the economic benefits of investments in information technology items. We will call such an approach a *functionalised* approach.

In the case of the banking industry, for instance, it is necessary to distinguish between the different *functions* of its production process. Mark distinguished three of them. For the Netherlands, it is appropriate to make a distinction between at least four or five different functions: (*1*) the intermediating function in the payment system; (*2*) activities such as participations and financing carried out within the so-called 'assets management' function; (*3*) operations to acquire financial means such as savings and deposits are reckoned among the 'liabilities management' function; (*4*) other financial services such as issuing shares and stock-jobbing; and, finally, (*5*) services originally not belonging to the banking profession, such as acting on behalf of a travel-agency.

The investments in information technology items and accordingly the relationships between capital and labour inputs and production volume output differ a lot among the five different functions mentioned. In carrying out the 'assets and liabilities management' functions, for instance, information technology items are scarcely used as far as the Dutch banking industry is concerned. However, a big part of value added is related to these two functions. The fourth function encompasses the activities carried out in the banks' trade rooms and technological change in the sense of investments is very impressive. In this field, however, the input of labour is very low if compared with other functions in the banking industry. A big part of employment is undoubtedly associated with the intermediating function in the *payment system*. This is the function where the process of technological change in Dutch banking started.

### *2.4 Trajectories and employment in the payment system*

Regarding the history of technological change, the concept of a *technological trajectory* as developed by Nelson and Winter (1982) is highly relevant. Trajectories in the banking industry of various countries differ according to the prevailing payment system. It is generally accepted to make a distinction between countries which have a 'cheque' system and others having a 'giro' system (De Wit, 1989). Technological developments in countries such as the

Netherlands with a '*giro*' payment system started with the automation of, first, the account records of clients in the 1960s, and second, of 'giro' transfers. This was realised by large investments in mainframe computer systems at central back offices of banks in connection with the development and introduction of new 'regular' payment instruments such as giro salary accounts and automated debits and credits. In contrast, in 'cheque' countries such as the United States, one was less able to develop comparable automated instruments for periodical payments such as salaries, mortgage payments and monthly rents. As a consequence, credit cards, cash dispensers and corporate electronic banking systems were introduced earlier and implemented on a much larger scale than in giro countries.

Returning to technological development in the Dutch banking industry, it is possible to characterise the starting point of the trajectory of information technology as the automation of the intermediating function in the payment system at central back offices. The trajectory developed itself in the field of this banking function at the central level up to 1980. Only at the end of the 1970s was the trajectory enlarged to encompass the local level of the payment function as well. Investments associated with this phase of the trajectory consisted of back-office terminals introduced from the end of the 1970s and front-office terminals at local bank offices from the beginning of the 1980s. Cash dispensers were introduced starting from the second half of the 1980s. Investments in technologies related to the 'assets and liabilities management' are expected to materialise in the 1990s.

The lesson to be learned from a historical analysis of technological change such as the one carried out above is that an economic examination in terms of productivity growth in the 1960s, 1970s and early 1980s should mainly be addressed towards the relationship in the field of the payment system at the central level; at least as it tends to relate productivity growth to investments in information technologies. If one realises that around 35 per cent of total employment in Dutch banking is located at central offices and furthermore that only a part of this 35 per cent is related to the processing of payment transfers, it is clear that a large part of banking employment in the 1960s and 1970s was untouched by the process of technological change. The conclusion from this is straightforward and would be a recommendation for case-studies within a sector. In the banking sector, such a case-study may be found in the case of automated clearing houses and in the computer centres of the central back offices of commercial banks. Such an approach based on the right relationship between inputs and outputs would facilitate the analysis of the economic benefits of investments in information technologies.

With respect to the impact of technological change in employment, it seems appropriate to make a distinction between at least two categories of labour. A first category may be characterised by its substitution through other production factors as a consequence of technological change. Within this category employees may be classified as those who carry out routine work which is affected by the new technologies, for instance data input. The other category may consist of employees including, for instance, managers, computer specialists and software engineers, who are complementary to new technologies.

Such a division could be applied to the Bank Giro Centrale (BGC), the

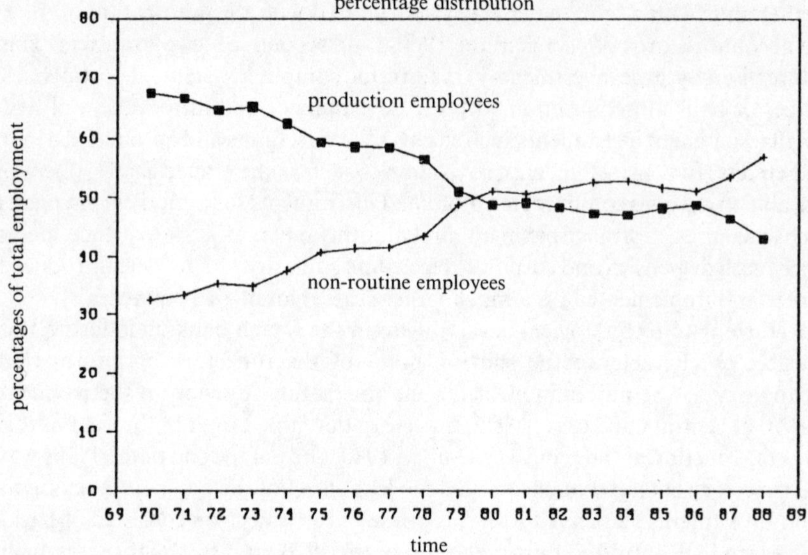


**Figure 1** Employment at BGC

automated clearing house of the commercial and retail banks in the Netherlands. In contrast with the traditional employment statistics of the whole banking sector (which classify 75 per cent of total employment as 'banking employees'), it is possible to distinguish between the two categories of labour at the level of an individual banking organisation such as the BGC. The employees occupied with data-input, data-processing, data-output and the electronic control of transfers are classified as production employees. Managers, staff personnel, software specialists and manual services employees are included in the other category of non-routine employees as defined above. Figure 1 illustrates the development of the relative shares of both categories of labour at the BGC during the 1970s and 1980s. It shows a systematic decline in the share of production employees and a steady increase in the share of non-routine employees. An analytical framework, however, is needed to examine the developments of both categories in relation to each other, the expenditures on information technology items and production volume. Such a framework is developed in the next section.

## 3. Modelling

### *3.1 Introduction*

The last section illustrated the new stylised facts of the services sector, in the sense of rapid technological change measured in terms of expenditures on information technology and a changing structure of the workforce. To assess the merits of these investments, however, it was argued that an analytical framework is needed to examine total factor productivity growth as well as the character of the process of technological change. Because the new information technologies are related to specific parts of the production processes in sectors, it was also argued that such a framework should not be applied to conventional sector statistics but to well chosen (parts of) companies. In the case of the Dutch financial industry, the automated clearing house of the Dutch commercial and retail banks called 'Bank Giro Centrale' (BGC) seems to be a good case in point. This section will be addressed towards the development of the analytical framework, while the application of the model to the case of the BGC will be carried out in Section 4.

The present section is divided into four further parts. First (3.2) we will review briefly the traditional models of technical change, which were mainly concerned with the share of production factors in total income. Next in Section 3.3, we will pay attention to the work of Salter, who was interested in the effects of technological change on the development of factor requirements in production. His model will be modified in two ways. The first adjustment will be described in Section 3.4 and is based on specific characteristics of technological change ruling out the possibility of substitution between factors of production. Finally, in Section 3.5 we work out the second adaptation consisting of the inclusion of more than two factors into the production function, making it possible to distinguish between the effects of technological change on different categories of employees.

### *3.2 Traditional Models*

Production functions have been used by many economists to analyse different aspects of technological change. An important distinction can be made between analyses of changes in the income distribution and studies directed towards the growth of productivity. Regarding the effects of technological change on the *shares of production factors in income* we can refer to the work of Hicks and Robinson. They were mainly interested in the change of factor shares in total national income. Consequently, they defined labour and capital saving biases of technological change within the framework of a production function as relative changes in marginal products assuming constant factor proportions. A relative rise in the marginal product of capital, for instance, would under the condition of constant factor proportions imply a labour-saving process of technological change.

Harrod's famous concept of neutral technological change was defined within the same tradition of growth theory and the distribution of national income: 'I define a neutral advance as one which, at a constant rate of interest, does not disturb the value of the capital coefficient. A stream of inventions, which are neutral as defined, will — provided that the rate of interest is unchanged — leave the distribution of the total national product as between labour and capital unchanged' (Harrod, 1948, p. 23).

Salter (1969, p. 32) clearly states that an analysis of the impact of technological change on *factor productivities* should 'reverse the procedure by assuming the marginal products constant and examine the effects on factor requirements'. Within such a framework, the *rate* of technological change can be defined as 'the relative change in total unit costs when the techniques in each period are those which would minimise unit costs when factor prices are constant', while the possible *biases* of technological change may be defined as 'the relative change in the ratio of two production factors when relative factor prices are constant'. The reverse of the change in total unit costs (total input costs per unit of output) is, of course, the rise in total product per unit of inputs. In that case the rate of technological change is often formulated as total factor productivity growth as it was originally developed in Solow's famous paper on technological change in the United States (Solow, 1957).

The models briefly discussed will not meet the goal of our own research. First, we do not intend to examine the effects of technological change on the distribution of income but on the unit requirements of production factors. Consequently, the approaches developed by Hicks, Robinson and Harrod are not directly applicable in the present research, which focuses on productivity and requirements of labour. Second, the level of our analysis is not the national economy but a specific sector or even the micro level of a representative company or firm. This observation is one more reason to reject the model of Harrod, who is assuming a fixed capital coefficient. The assumption of the fixed capital coefficient was derived from the notion that the economy consisted of two sectors: one capital goods producing sector; and another sector using the newly developed capital goods. The capital goods producing sector was assumed to develop new knowledge and to produce new capital goods setting into motion a process of technological change reflected in decreasing unit costs. Furthermore, the capital goods using industries would substitute capital goods for labour as a consequence of the relative price decrease of the capital goods. At the aggregate level the assumption was made that the substitution effect was just big enough to compensate for the initial savings in capital as a consequence of technological change. According to Salter (1969, p. 41), at the sectoral level, however, it is essential to think in two steps; the technological advances actually taking place in each and different industries, on the one hand; and the cheapening of capital goods which is common to all industries, on the other.

We conclude that Salter's approach is the most promising in the sense that his model is directed towards examining the development of unit factor requirements at a sectoral level. It can be shown that this is the reversal of the original Hicks model, applied at a sectoral level instead of the national aggregate, in the sense that it is not the factor proportions which are assumed constant but the marginal products of the factors of production (Salter, p. 33). The definition of the character of technological change has, as far as neutrality is concerned, the same implications as the definition used by Hicks. Hicks neutral advance requires an unchanged ratio of marginal products when factor proportions are constant, and the definition of Salter implies that factor proportions are unchanged when the ratio of marginal products is constant. The character of technological change in terms of labour-saving or capital-saving imply the same directions of the process but differ in the measure of their extent. Salter's model measures the bias in terms of the relative changes in the factor-intensity assuming constant marginal factor products, while Hicks is interested in relative changes of factor shares in income assuming constant factor proportions. Furthermore, the rate of technological change is the same as in the model of both Hicks and Solow, although it is not measured in terms of total factor productivity growth but in terms of a decrease of total unit factor costs. In the next section Salter's model will be presented in a more formalised way.

### *3.3 Salter's Model*

Salter's model is based on a neo-classical production function with two factors of production: capital and labour. Within this framework, the *rate* of technological change ($T^{\circ}$) is defined as 'the relative change in total unit costs when the techniques in each period are those which would minimise unit costs when factor prices are constant'. Furthermore, he is interested in the explanation of the relative changes of unit factor requirements of labour and capital. He shows that these relative changes can be explained by the rate of technological change ($T^{\circ}$) as defined above, the *bias* of technological change defined as 'the relative change in the ratio of two production factors when relative factor prices are constant' and the *elasticity of substitution* between the factors of production. In a formalised manner (for a detailed derivation of the relationships, see Salter, 1969, pp. 30–2 and 46–7, as well as Section 3.5 of the present chapter):

$$L^{\circ} = (dL / dt) \,.\, (1 / L) = T^{\circ} - a \,.\, B^{\circ}_{l} + s \,.\, a \,.\, (1 / p^{\circ}) \quad (1)$$

$$C^{\circ} = (dC / dt) \,.\, (1 / C) = T^{\circ} + (1 - a) \,.\, B^{\circ}_{l} + s \,.\, (1 - a) \,.\, p^{\circ} \quad (2)$$

$$T^{\circ} = \frac{(dL / dt) \,.\, w + (dC / dt) \,.\, g}{L \,.\, w + C \,.\, g} \quad (3)$$

$$B^{\circ}_{l} = d\,(C / L) / dt \,.\, (L / C) \quad (4)$$

in which:

| | | |
|---|---|---|
| $L^{\circ}$, $C^{\circ}$ | : | proportionate growth rates of labour and capital requirements per unit of output; |
| $T^{\circ}$ | : | proportionate growth rate of technological change, i.e. relative change in total unit costs; |
| $B^{\circ}_{l}$ | : | represents the factor-bias of technological change, implying capital-saving technological change if $B^{\circ}_{l} < 0$, neutral technological change if $B^{\circ}_{l} = 0$ and labour-saving technological change if $B^{\circ}_{l} > 0$; |
| $a$ | : | the share of capital costs in total costs or: $C.g / (L.w + C.g)$; |
| s | : | the elasticity of substitution between capital and labour; |
| $p^{\circ}$ | : | proportionate growth rate of the relative factor price ratio. |

Salter (1969, p. 33) states that the pace and character of technical advance varies markedly from one industry to the other, and refers to Weintraub to illustrate the possibility of the simultaneous occurrence of both labour-saving and capital-saving processes of technological change within one and the same industry. Weintraub (1939) analysed the assembly-line principle in car manufacturing, which was often regarded as a highly labour-saving innovation. In overall 'biased' terms this was, according to Weintraub, undoubtedly true, but at the same time it was undoubtedly true that technological change did save in floor-space, stocks and works-in-progress and the absolute increase in capital productivity might well have been in the same order as that of labour productivity increase.

It is our goal of research to examine the influence of technological change on both the unit factor requirements of capital and different categories of labour. Consequently, the production function of Salter will be extended to include more factors. But first, we will argue why the substitution effects responding to changes in relative factor prices specified in the equations (1) and (2) above are of no concern if the character of technological change itself is taken into account.

### *3.4 The Leontief Production Function*

In their discussion of the issue of technological change and factor substitution, Freeman and Soete follow the ideas of Atkinson and Stiglitz (1969) regarding 'localised' technological change. Atkinson and Stiglitz first assess that the neo-classical growth accounting studies which were undertaken in the 1950s and the 1960s assumed that technological progress could be represented by an outward shift of the entire production function. In other words, technical advance was assumed to raise output per head for all possible techniques. They continue: 'the advocates of this approach seem to have forgotten the origins of the neo-classical production function: as the number of production processes increases (in an activity analysis model), the production possibilities can be more and more closely approximated by a smooth, differentiable curve . . . If the effect of technological advance is to improve one technique of production but not other techniques of producing the same product, then the resulting change in the production function is represented by an outward movement at one point and not a general shift' (1969, p. 573). If technical progress is 'localised' to one technique, both production factors capital and labour can augment and this is likely to result in reductions in the short-term elasticity of substitution. In other words, the 'smooth' neo-classical production function is gradually being transformed into a classical Leontief production function with fixed technical coefficients in which the role of the relative factor ratio is ruled out (as far as the short-term is concerned, see Figure 2).

In our opinion, technological change in a service industry such as banking is 'localised'. All banks operating in the field of the payment system introduced more or less the same techniques, i.e. mainframe computer technologies delivered by either IBM or Burroughs (Unisys). Because of the rapid developments in computer technology itself, the systems have been replaced every five years and banks have proved to be 'supplier-sticky', resulting in a process of technological change which is more or less 'irreversible' and limiting the substitution possibilities between capital and labour significantly.

According to Atkinson and Stiglitz (1969, p. 576) 'the effect of localisation can be so strong that the advanced techniques dominate the less capital-intensive ones, requiring both less labour and less capital'. The interesting point of analysis in such a case is that any substitution between capital and labour as explained by changes in the relative factor price ratio is no longer relevant and the direction and rate of technological change within a historical and institutional framework become the main explanatory factors. Atkinson and Stiglitz (1969, p. 577) give

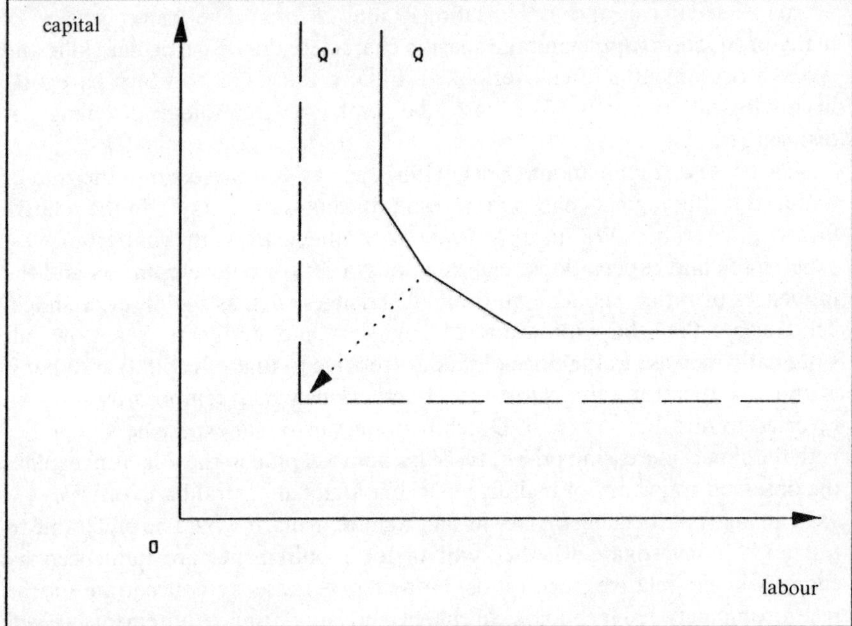


**Figure 2** The Leontief Production Function

the example of an economy which is in a long-run equilibrium using a relatively labour-intensive technique when suddenly a plague wipes out a large proportion of the labour force, so that wages rise and a more capital-intensive technique becomes attractive and will be adopted. Technological change is localised now into that technique and it is unlikely that the economy will ever return to the more labour-intensive technique.

A comparable development occurred in banking. During the 1950s and the early 1960s, the economy was growing very fast and labour was scarce. In addition, the demand for profitable loans was large. For banks it was interesting to enlarge the period during which they had the money of clients at their disposal by the introduction of salary and current accounts in order to acquire relatively cheap funds. The different processes together, i.e. those of relative shortages of labour supply and cheap money funds and the rapid increase in the volume of payment transactions, resulted in the introduction of mainframe computer technologies at central bank offices as described in the last section. Once these technologies were introduced, they proved to be the main 'technological trajectory' during the 1960s and 1970s which gave shape to the direction and rate of technological change in banking.

The same arguments in favour of 'localised' technological change have been brought to the fore by Freeman and Soete (1987, pp. 42, 43). They argue that technological change predominantly results from the accumulation of technical knowledge about specific techniques within firms. Further, they argue that a 'smooth' production function ignores the interrelatedness and complementarities of many technical and organisational innovations. The heterogeneity of many production inputs and the specific characteristics of particular skills and types of equipment is often overlooked. Finally, the extent to which particular 'technological trajectories' dictate the path of technological change is disregarded.

Nevertheless, Freeman and Soete (1987, pp. 44–8) do recognise the role of 'induced technological change' as a result of long-term changes in the relative factor price ratio. Within this framework they stress the importance of experiences and expectations regarding long-term price developments and the influences of rather sudden and dramatic changes such as the oil price shock. They argue that the expectation of engineers and designers is a slow but systematic increase in the price of labour (relative to that of capital) and that it is unlikely, therefore, that short-term fluctuations would temporarily result in reversed trends. In the case of Dutch banking, historical experiences, together with the slowly increasing price ratio of labour to capital in the long run, explain the observed trajectory of mainframe technologies at central bank offices.

In an analysis focusing on labour unit requirements, it would be preferable to test and to investigate whether unit factor requirements are influenced by changes in the relative price ratios. However, because we concentrate on the relationship between technological change and labour unit requirements we will assume a Leontief production function and disregard possible substitution effects. Such an approach makes it easier to highlight the interrelationships between more than just two production factors. Furthermore, the assumption seems realistic bearing in mind that technological change in the payment system is 'localised'. Consequently, the relationships between factor productivities and the direction and rate of technological change according to the Salter-model in Section 3.3 above boils down to the following (in which the influence of the relative factor price ratio is cancelled out):

$$L^{\circ} = T^{\circ} - a \,.\, B^{\circ}_{l} \tag{5}$$

$$C^{\circ} = T^{\circ} + (1 - a) \,.\, B^{\circ}_{l} \tag{6}$$

in which:

$L^{\circ}, C^{\circ}$ : proportionate growth rates of labour and capital requirements per unit of output;

$T^{\circ}$ : proportionate growth rate of technological change, i.e. relative change in total unit costs;

$B^{\circ}_{l}$ : represents the factor-bias of technological change, implying capital-saving technological change if $B^{\circ}_{l} < 0$, neutral technological change if $B^{\circ}_{l} = 0$ and labour-saving technological change if $B^{\circ}_{l} > 0$;

$a$ : the share of capital costs in total costs or: C.g / (L.w + C.g).

### *3.5 The Adapted Model*

In this section, we will adapt Salter's basic model to include as many production factors as necessary. The basic definitions of the rate and the bias of technological change, however, remain the same as defined in Section 3.3. Nevertheless, equations (1), (2) and (3) will be reformulated in order to include the total number of production factors specified. The number of biases, i.e. the relative changes in factor intensities formalised as in equation (4), will increase to a total of [½.n.(n–1)] in case of n factors. Furthermore, the substitution effect is cancelled according to equations (5) and (6) above.

We start the formal derivation by replacing Salter's original production function with a loosely defined function incorporating (n–j+1) production factors $F_i$.

$$Q = f\,(F_j, F_{j+1}, \ldots, F_k, \ldots, F_{n-1}, F_n) \tag{7}$$

Technological change T° measured as the relative change in total unit costs can accordingly be formulated as the sum of changes in n factors $F_i$, multiplied by their respective original marginal products $f_i$ and divided by the original total costs:

$$T^\circ = \frac{\sum_{i=j..n} (\, dF_i \,/\, dt) \,.\, f_i}{\sum_{i=j..n} F_i \,.\, f_i} \tag{8}$$

In the same way a bias $B^\circ_{ij}$ may be defined as the relative change in the ratio between two production factors $F_i$ and $F_j$ when relative factor prices $f_i$ and $f_j$ are constant.

$$B^\circ_{ij} = \frac{d(\, F_i \,/\, F_j\,) \,/\, dt}{F_i \,/\, F_j} = \frac{dF_i \,/\, dt}{F_i} - \frac{dF_j \,/\, dt}{F_j} = \tag{9}$$

$$F^\circ_i - F^\circ_j$$

$$F^\circ_i = B^\circ_{ij} + F^\circ_j \tag{10}$$

$$dF_i \,/\, dt = [\, B^\circ_{ij} + F^\circ_j \,] \,.\, F_i \tag{11}$$

rewriting equation (8):

$$T^\circ \,.\, \textstyle\sum_{i=j..n} F_i \,.\, f_i = \sum_{i=j..n} (\, dF_i \,/\, dt\,) \,.\, f_i \tag{12}$$

$$T^\circ \,.\, \textstyle\sum_{i=j..n} F_i \,.\, f_i = dF_j \,/\, dt \,.\, f_j + \sum_{i=j+1..n} (\, dF_i \,/\, dt) \,.\, f_i \tag{13}$$

rearranging terms yields:

$$(dF_j/dt).f_j = T^\circ \,.\, \textstyle\sum_{i=j..n} F_i \,.\, f_i - \sum_{i=j+1..n} (dF_i \,/\, dt) \,.\, f_i \tag{14}$$

substituting (10) in (14):

$$(dF_j/dt).f_j = T^\circ \,.\, \textstyle\sum_{i=j..n} F_i \,.\, f_i - \sum_{i=j+1..n} [\, (B^\circ_{ij} + F^\circ_j) \,.\, F_i \,] \,.\, f_i \tag{15}$$

$$(dF_j/dt).f_j = T^\circ \,.\textstyle\sum_{i=j..n} F_i.f_i - \sum_{i=j+1..n} B^\circ_{ij}.F_i.f_i - \sum_{i=j+1..n} [(F_i.f_i)/F_j].(dF_j/dt) \tag{16}$$

rearranging terms:

$$(dF_j/dt).f_j + \textstyle\sum_{i=j+1..n} [(F_i.f_i)/F_j] \,.\, (dF_j/dt) = T^\circ.\sum_{i=j..n} F_i.f_i - \sum_{i=j+1..n} B^\circ_{ij}.F_i.f_i \tag{17}$$

$$(dF_j/dt) \,.\, [\, f_j + \textstyle\sum_{i=j+1..n} (F_i.f_i)/F_j \,] = T^\circ \,.\, \textstyle\sum_{i=j..n} F_i.f_i - \textstyle\sum_{i=j+1..n} B^\circ_{ij} \,.F_i.f_i \qquad (18)$$

dividing by $\sum_{i=j..n} F_i \,.\, f_i$ yields:

$$(dF_j/dt) \,.\, (1/F_j) = T^\circ - \textstyle\sum_{i=j+1..n} [(F_i.f_i) \,/\, (\textstyle\sum_{i=j..n} F_i.f_i)] \,.\, B^\circ_{ij} \qquad (19)$$

$$F^\circ_j = (dF_j/dt) \,.\, (1/F_j) = T^\circ + \textstyle\sum_{i=j+1..n} [\, (F_i.f_i) \,/\, (\textstyle\sum_{i=j..n} F_i.f_i) \,] \,.\, [B^\circ_{ji}] \qquad (20)$$

From equations (19) and (20), it is clear that the relative changes in *unit requirements* of an arbitrarily chosen production factor can be expressed in terms of the rate of technological change and a weighted sum of its biases with the other factors of production. This model will be used in the next section to analyse the process of technological change and its effects on different categories of labour at the BGC.

## 4. Banking

### *4.1 Introduction*

In Section 2.2 it was explained that the financial services industry was one of the areas in the services sector most susceptible to applications of information technology. Furthermore, in Section 2.3. we made clear that in assessing the benefits of new information technologies it is necessary to analyse specific parts of an industry in order to relate the right outputs with relevant inputs. It was argued that an automated clearing house could serve as a representative case-study of the banking sector. Consequently, the model developed in the last section will be applied to the case of the BGC, the automated clearing house of the Dutch commercial and retail banks.

The BGC was established in 1967 as a response of the banks to the central computer centres of the Postbank, at that time the government owned Postal Giro Services. The period up to 1970 should be viewed as the take-off, while the re-organisation of the processing of payment transfers from the associated banks to the BGC covered the period up to the mid-1970s. Our examination especially focuses therefore, on the period from 1975. Within this period a few sub-periods can be distinguished in which main changes of information technologies took place. These changes were associated with replacements of the mainframe computer systems and the buildings in which the clearing house was operating. In general the mainframe systems were replaced every five years, starting at the beginning of the 1970s, and successive replacements took place around 1976, 1981 and 1985. In addition, a new production centre was opened in 1978 which was also accompanied by a partial replacement of mainframe computer systems. Furthermore, it is worth while to notice that during the whole lifetime of the BGC the information capital stock was continuously adapted in terms of memories, processing units and peripheral equipment in order to be able to process the transfers offered by the member banks. Finally, the development and introduction of mini- and micro computers as well as electronic payment instruments were implemented during the course of the 1980s introducing a new element to the whole production process.

Before analysing the process of technological change in more detail, we will present some overall developments at the BGC and pay attention to the composition of the information technologies used. Finally, the process itself in terms of rate and character is addressed.

### *4.2 Output, Employment and Information Technology at the BGC*

The general economic development of the BGC in terms of production volume, employment and other factor inputs such as housing and information technologies is presented in Figure 3. This shows increasing levels of production volume and factor inputs. The rate of growth in the case of production volume, measured in terms of different numbers of payment transfers and guaranteed cheques, however, is much higher than the corresponding growth rates of labour inputs and total factor inputs. The volume of production grew from an index of less than 10 in 1970 (1975 = 100) to 320 in 1987, while labour inputs and total factor inputs grew from around 20 to around 150 during the same period. The average annual growth rates related to the period from 1975 can be calculated at 10 per cent and 3.6 per cent, for production volume and inputs respectively. Measuring total factor and labour productivity growth as the difference between growth rates in production volume and inputs, a resulting total factor productivity growth rate of 6.4 per cent is obtained, while the partial labour productivity growth is even higher. These first indicators seem to be more in line

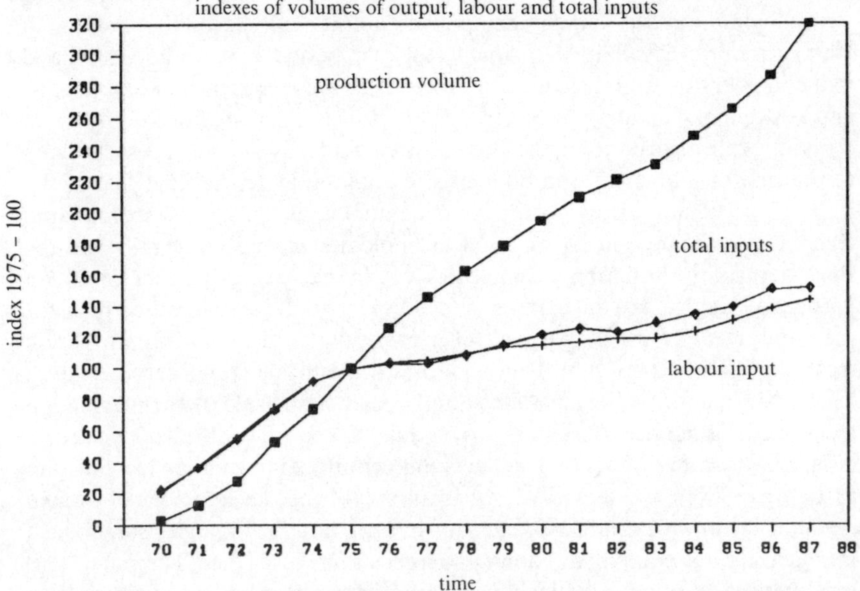


**Figure 3** General economic development at the BGC

with the results obtained by Mark than those of Kendrick and Roach as reported in Section 2.3.

To get further insights into the process of technological change, it should be noticed that labour inputs as well as other inputs such as those related to information technologies consist of different components. With reference to Section 2.3, it is worth recalling that the two categories of labour at the BGC, i.e. those of payment production employees and those of non-routine employees, showed contrary developments in terms of their shares in total employment. The analytical framework of the last section is needed to show the development of productivity of each factor and the extent of substitution between the two factors. Such an analysis will be carried out in Section 4.3.

In the same section, we will analyse the development of information technology productivities and possible processes of substitution between the two different categories of labour and information technologies. First of all, however, we pay attention to the *composition of information technologies*. Traditionally, growth accounting studies concentrate on the relationships between output and two inputs, i.e. labour and capital. Although the services of capital should be used in the production function, one is used to indexes of the capital stock based on the assumption that the capital services are closely related to the stock. However, in the case of service industries three observations make such an approach less valid.

First, as was shown by Roach (1988) in the case of the United States, many items of information technology are not owned by the user industry but leased from other firms. Calculations based on the traditional capital stock, therefore, will under-estimate the inputs of capital services into the production process. The relative importance of information technology in the case of services, as was shown in Section 2.2, clearly underlines the inclusion of leasing of information technologies into the capital component considered. Secondly, traditional vintage models calculate the capital stock at period t as additions to and reductions in the capital stock at period (t-1) as a consequence of investments and scrapping. The amount of scrapped capital is calculated according to an economic rule incorporating the idea of 'quasi' rents. Such a rule does not apply to the case of information technologies in banking in general and that of the BGC in particular. Among other considerations, as a consequence of rapid technological change in information technologies themselves, banks and the BGC replace information technologies every five years, whether or not a particular vintage is still earning a positive rent. It seems rather realistic, therefore, to associate 20 per cent of a particular investment in information technologies for a period of five years to the amount of capital services input. Third, the amount of maintenance and repair associated with information technologies is not necessarily stable over time. It is conceivable, that it decreases because of learning effects both at the manufacturer of information technologies as well as within the user firm. Figure 4 shows the development of annual amounts spent on the total category of information technologies sub-divided into leasing of equipment, annual depreciation of owned computer and communication equipment including communication costs and finally, costs associated with maintenance and repair.

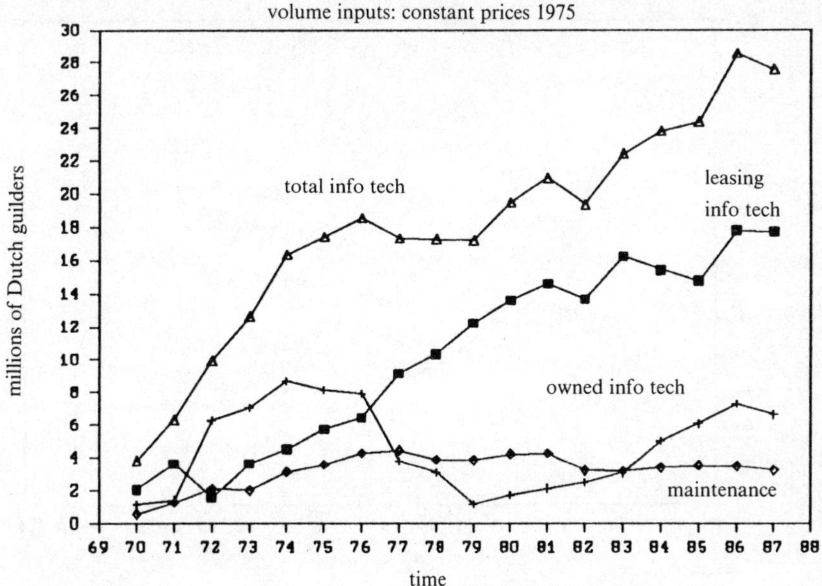


**Figure 4** Expenditures on information technology

Total information technologies input at the BGC have increased rapidly over the last seventeen years. The most rapid increase was in the period up to 1977, with an annual growth rate of around 24 per cent, a more-or-less constant level at the end of the 1970s and a further increase from 1979 up to 1987, with a rate of 6.5 per cent annually. A few remarkable further conclusions can be drawn. First, the share of leasing in the total of information technology capital services is considerable and should certainly not be ignored. Second, the share of owned equipment and communication costs is fluctuating. In the first half of the 1970s capital services from owned equipment were far more important than those of leased equipment. After 1976 this changed as a consequence of different policies at the BGC including preferences for leasing of mainframe computer equipment. From the beginning of the 1980s, however, the input of owned equipment and communication is rising again. This is mainly a result of investments in mini- and microcomputer systems as well as communication costs associated with systems of electronic payment at the point-of-sale and cash dispensers. Finally, the amount of maintenance and repair is more or less stable but declining if measured in terms of percentages of total information technology inputs.

### *4.3 Rate and character of technological change at the BGC*

We will calculate the *rate* of technological change at the BGC, according to equation (8) of the adapted model. Because our focus of analysis is on technological change and employment, we include as factors of production two categories of labour (production employees and non-routine employees), three components of information technologies (leasing, depreciation costs of own

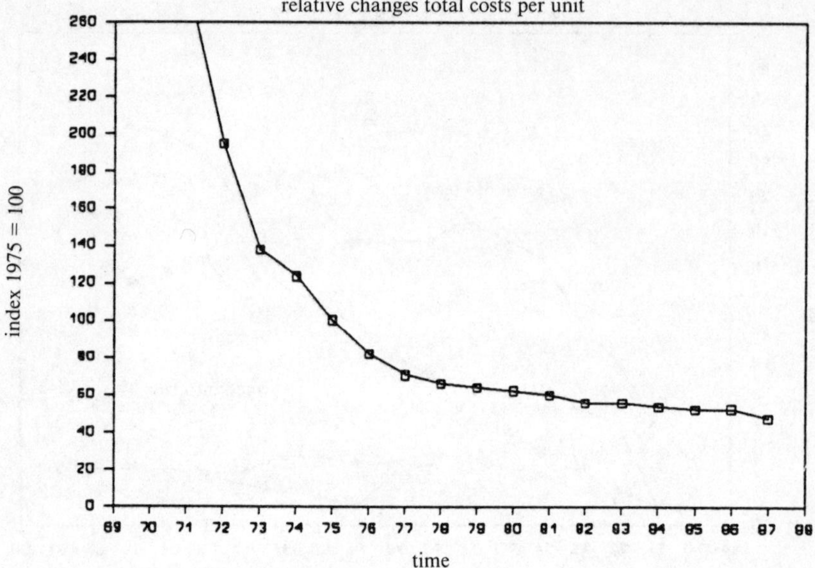


**Figure 5** Rate of technological change

equipment including communication costs and maintenance and repair) and housing costs. Other inputs having less association with the technological characteristics of the production process are not included. The rate thus calculated is presented in Figure 5. This shows a high rate of technological change up to the end of the 1970s, with a decrease in per unit costs of 35 per cent per annum during the period 1970–78; however, this high rate is mainly a consequence of idle capacity at the start of the BGC and scale economies. After 1978 the rate decreased to a level of 3.8 per cent annually, which is still considerable. This boils down to a rate of 6.4 per cent after 1975, as was already noted in Section 4.2.

The *nature* of technological change at the BGC can be characterised into much more detail than just labour or capital saving. Using equation (9) of the adapted model, we may analyse the relative change in any ratio of two production factors. Furthermore, it is possible to define a total bias of a specific production factor as the relative change of that factor in relation to all inputs per unit of output. The different developments in total biases of the different production factors can be combined in one figure illustrating the percentage share of each factor in total inputs per unit of output. This has been done in Figure 6.

Leaving aside the first two years of the 1970s, Figure 6 shows a few noteworthy and steady trends in the process of technological change. The inputs of non-routine employees and information technologies per unit of output are increasing, while those of production employees and buildings are decreasing. In more detail, the input of non-routine employees has increased from 36 per cent in 1973 to 42 per cent in 1987, while the information technologies input increased from 30 per cent to 38 per cent. The downward trend in the other two factors can

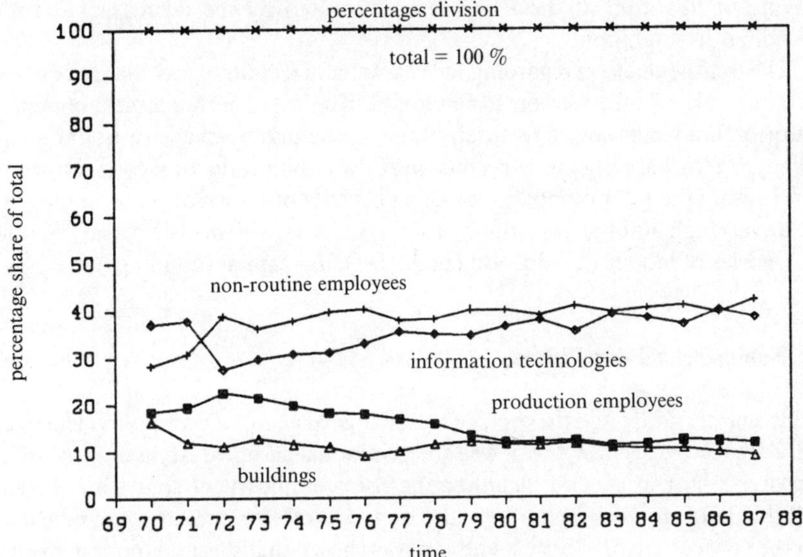


**Figure 6** Inputs per unit of output at the BGC

be estimated at a decrease from 21 per cent to 11 per cent in the case of production employees, and from 13 per cent to 9 per cent in the case of buildings.

A few further points are worth noticing. First, when we add both categories of labour into one factor of production, as traditional analysis is used to, a slightly downward trend, from 57 per cent to 53 per cent of total labour inputs per unit of output, would appear. However, as has been shown, the underlying trends of different categories of labour consist of steady but contrary developments. Consequently, we prefer to characterise this process of technological change not as labour-saving but as saving in production employees, on the one hand, and creating non-routine employees, on the other.

Second, regarding the different components of capital inputs a comparable qualification can be made. Instead of capital-augmenting (or labour-saving), we prefer to use the terms of an information technologies augmenting and a building saving process of technological change.

Third, regarding the technological relationships between information technology and labour, the process could be characterised as saving in both categories of labour (as could be shown by the individual biases) because the increase in the share of information technology is more rapid than in the case of both categories of labour (production and non-routine employees).

Finally, the rising share of information technologies and the decreasing share of buildings is a clear confirmation of Weintraub's propositions regarding capital-saving quoted by Salter in Section 3.3. According to Weintraub the assembly-line in car manufacturing could have saved in floor-space, stocks and work-in-progress. Our analysis shows that the rapid growth in memories and processing speed of mainframe computer technologies did result in capital saving of this kind at the BGC as is illustrated by the decreasing share of buildings in total inputs.

The final conclusion regarding the character of technological change concerns the key role of information technologies. The rapid technological change in information technologies themselves acts as the major vehicle of technological change in the banking industry, pushing down unit production costs. However, in realising the full potential of possible benefits of information technologies a relatively high input of non-routine employees is needed, in order to save in other categories of labour (production employees) and capital (buildings).

## 5. Summary and conclusions

This chapter addressed the topic of technological change and employment in the services sector. Section 2 reviewed the traditional assumed 'stylised facts' of the services sector in general including the rising employment shares and lagging productivity growth. Technological change was often assumed not to take place in the services sector. Subsequently, it was shown that this assumption is out of date. In fact, the last thirty years showed a rapidly increasing and changing nature of the capital stock in the services sector, mainly as a consequence of information technologies.

Regarding the transformation of investments in technologies into productivity growth, however, several authors are rather pessimistic as far as the services sector is concerned. In our opinion the so-called 'productivity-paradox' could at least partially be solved by combining an economic analysis of technological change with a historical and institutional approach throwing light on the characteristics of the technologies themselves. Such an analysis shows, as in the case of financial services, that new technologies are used in specific parts of production processes of sectors or firms. Consequently, traditional productivity analyses starting from data applying to the whole branch can measure the benefits of information technology in relation to other branches only in a rough way. A 'functionalised' approach would select a representative firm or a specific (part of a) production process. In the case of Dutch banking the payment system appeared to be the function in which technological change — as far as computer and communication technologies are concerned — started and where most developments so far actually took place. Subsequently, this chapter selected the Bank Giro Centrale (BGC) as representative of the payment system in the Netherlands.

Next, in Section 3 traditional models of technological change were reviewed. From a perspective of the relationship between technological change and the level of employment, the original model of Salter appeared to be the most promising. However, the model did also incorporate two characteristics which were not suited for own research: first, substitution effects as a consequence of changes in relative prices and second, the specification of two factors of production only. Both characteristics were considered inappropriate for the present analysis. Principally on the basis of arguments expressed by first, Atkinson and Stiglitz, and second, Freeman and Soete, the substitution mechanism in reaction to relative price changes was left out of the analysis. Furthermore, the changes in the composition of both labour and capital stimulated the choice of a model including more factors of production. The original Salter model was adapted in both respects.

Against the background of Sections 2 and 3, an analysis of the process of technological change and employment at the BGC was carried out in Section 4. The main conclusions can be formulated as follows. First, traditional (vintage) growth accounting studies based on the capital stock are less suited if a major part of information technologies has been leased. According to our calculations, the leasing of information technologies in the case of the BGC covers 50 per cent up to 75 per cent of total information technology services. Second, the application of information technologies has been translated into strong decreases of total costs per payment transfer. In other words, productivity growth at the BGC was very strong indeed, and higher than most parts of manufacturing during the last ten to fifteen years. Third, regarding the character of technological change, it is possible to describe the process in much more detail than just labour- or capital-saving. In the case of the BGC it was shown that the process of technological change saved in the two groups of labour specified, i.e. those of production employees, on the one hand, and non-routine employees, on the other, if analysed in relation to the inputs of information technologies. However, if studied in relation to all other inputs the bias of non-routine employees was positive. Fourth, next to the savings in labour, information technologies set into motion a process of saving in other categories of capital as well. This was illustrated by the savings in buildings in the case of the BGC.

Although the BGC is not wholly representative of the commercial services sector in general, we believe that the approach is applicable to other parts of the services sector as well. A few necessary conditions, however, should be fulfilled. First, the process of technological change in a services branch should be analysed from a historical and institutional perspective in order to determine those parts or function(s) of the branch which are representative for the implementation of new information technologies. Second, within such a function a firm or a production process should be found which provides the necessary data. Third, having established that a process of technological change in services can have quite contrary impacts on different categories of labour, it is recommended to distinguish at least between complementary and substitutionary labour with respect to the new technologies. Finally, one should investigate the behaviour regarding the depreciation schemes in the case of investment in technologies and whether or not information technologies are leased to a considerable extent.

**Acknowledgement**

The author wishes to thank the members of MERIT's productivity group (Paul Diederen, René Kemp, Huub Meijers, Luc Soete, Bart Verspagen and Adriaan van Zon) and Frits Prakke for comments on earlier drafts and ideas; the analysis carried out, the results obtained and any errors or omissions remain the full responsibility of the present author; furthermore, the enthusiastic and cordial co-operation of the BGC is gratefully acknowledged.

## References

Atkinson, A.B. and J.E. Stiglitz (1969), 'A new view of technological change', *The Economic Journal*, 79, pp. 573–8.

Baily, M.N. and A.K. Chakrabarti (1988), 'Electronics and white-collar productivity', *Innovation and the Productivity Crisis*, Washington DC: the Bookings Institution, pp. 86–103.

Bank Giro Centrale (BGC) (1988), *Annual Reports 1970–1988*, Amsterdam: BGC.

Baumol, W.J. (1967), 'Macroeconomics of unbalanced growth: the anatomy of the urban crisis', *American Economic Review*, 49, pp. 415–26.

Dosi, G. (1984), *Technical Change and Industrial Transformation: the Theory and an Application to the Semiconductor Industry*, London: Macmillan.

Elfring, T. (1988), 'Service Employment in Advanced Economies: a Comparative Analysis of its Implications for Economic Growth', Ph.D. thesis, University of Groningen.

Freeman, C. and L. Soete (eds) (1987), *Technical Change and Full Employment*, Oxford: Basil Blackwell.

Fuchs, V.R. (1964), 'Productivity Trends in the Goods and Service Sectors, 1929–1961: a Preliminary Survey', *Occasional Paper 89*, New York: NBER.

Fuchs, V.R. (1965), 'The Growing Importance of the Service Industries', *Occasional Paper 96*, New York: NBER.

Fuchs, V.R. (1968), *The Service Economy*, New York: NBER.

Fuchs, V.R. (ed.) (1969), 'Production and Productivity in the Service Industries', *NBER Studies in Income and Wealth 34*, New York: NBER.

Gorman, J.A. (1969), 'Alternative measures of the real output and productivity of commercial banks', in Fuchs, V.R. (ed.), *Production and Productivity in the Service Industries*, New York: NBER.

Guy, K. (1987), 'The UK tertiary service sector', in Freeman, C. and L. Soete (eds), *Technical Change and Full Employment*, Oxford: Basil Blackwell, pp. 169–88.

Harrod, R.F. (1948), *Towards a Dynamic Economics: Some Recent Developments of Economic Theory and their Applications to Policy*, London: Macmillan.

Hicks, J.R. (1963), *The Theory of Wages* second edition, London: Macmillan.

Kendrick, J.W. (1988), 'Productivity in services', in Guile B.R. and J.B. Quinn (eds), *Technology in Services: Policies for Growth, Trade and Employment*, Washington, DC: National Academy Press, pp. 99–118.

Kimbel, D. (1973), *Computers and Telecommunications*, Paris: OECD.

Mark, J.A. (1988), 'Measuring productivity in services industries', in Guile, B.R. and J.B. Quinn, *Technology in Services: Policies for Growth, Trade and Employment*, Washington, DC: National Academy Press, 1988, pp. 139–60.

Nelson, R.R. and S.G. Winter (1982), *An Evolutionary Theory of Economic Change*, Cambridge, Mass.: The Belknap Press of Harvard University.

Roach, S.S. (1988), 'Technology and the services sector: America's hidden competitive challenge', Guile, B.R. and J.B. Quinn (eds), *Technology in Services: Policies for Growth, Trade and Employment*, Washington, DC: National Academy Press, pp. 118–39.

Robinson, J. (1937), 'The classification of inventions', *Review of Economic Studies*, 5, pp. 139–42.

Salter, W.E.G. (1969), *Productivity and Technical Change*, paperback edition, Cambridge: Cambridge University Press.

Solow, R. (1957), 'Technological change and the aggregate production function', *American Economic Review*, 39, pp. 312–20.

Weintraub, D. (1939), 'The effects of technological development on capital formation', *American Economic Review*, 29 (1), pt 2 supplement, pp. 15–32.

Wit, G.R. de (1989), 'Dutch banking: technological developments', in Muysken, J. (ed.), *Technological Change, Employment and Skill Formation in Dutch Banking*, Research Memorandum LIB/R/89/002, LIBER, Maastricht, University of Limburg, pp. 14–26.

# 6. Diffusion of information technology in banking: the Netherlands as an illustrative case[1]

*P.J.M. Diederen, R.P.M. Kemp, J. Muysken and G.R. de Wit*

## 1. Introduction

Since the 1970s, the diffusion of new information technologies has induced a process of structural change in Western economies. The financial sector has been one of the branches of economic activity that has benefited most from the development of computer and communication technologies. Since these technologies facilitate decentralization, they have a profound impact on the division of labour, within and between organizations, and on patterns of employment, like part-time and self-employment.

The goal of this chapter is to describe and to explain the diffusion of information technology in the Dutch banking industry. We start with a description of information technology and its diffusion in different branches of the Dutch economy, indicating the relatively high penetration of information technology in banking. In addition, an outline is given of the trajectory of technical change in banking. In Section 3 a theoretical framework is developed, putting the diffusion process in perspective. Section 4 specifies a model to analyse the diffusion of technology and Section 5 estimates the model using data from the Dutch banking industry. In Section 6 the main conclusions of the chapter are underlined and attention is paid to possible implications for policy measures.

## 2. Information technology

Inventions and innovations, the commercial application of inventions, are considered crucial elements in techno-economic development. Freeman distinguishes between four different categories of innovation: incremental innovations, radical innovations, technological systems and techno-economic paradigms.[2] Without going into details, it can be stated that the overall economic effects are lowest in the case of incremental innovations and highest, affecting all parts of the economy, in the case of techno-economic paradigms. Initially, the concept of a new techno-economic paradigm was developed by Perez and it has been associated with other related concepts such as those of 'clusters of innovation' developed by Schumpeter, 'technological trajectories' by Nelson and Winter and 'technological paradigms' by Dosi.[3] According to Perez, a new techno-economic paradigm satisfies at least the three following conditions: (1) rapidly falling relative costs during a considerable period of time; (2) a large and sufficient supply of the new systems and equipment; and (3) obvious possibilities for the application of the key elements of the new paradigm in both existing

**Table 1** Administrative IT – applications in the Netherlands, 1986

| *Sector* | *Total costs Fls, millions* | *Number of terminals* | *Costs per employee* | *Terminals per 1,000 employees* |
|---|---|---|---|---|
| Agriculture | 34 | 1 132 | 460 | 15.30 |
| Food | 300 | 5 784 | 2 050 | 39.62 |
| Chemicals | 733 | 14 627 | 5 820 | 116.09 |
| Engineering | 1 201 | 31 290 | 3 320 | 86.44 |
| Manufacturing | 2 711 | 65 213 | 3 270 | 78.76 |
| Construction | 172 | 4 589 | 500 | 13.38 |
| Trade | 1 897 | 49 956 | 2 110 | 55.45 |
| Transport/communication | 542 | 16 878 | 1 650 | 51.30 |
| Banking | 2 001 | 33 741 | 17 100 | 288 38 |
| Insurance | 942 | 21 316 | 15 700 | 355.27 |
| Commercial services | 6 090 | 108 944 | 8 510 | 152.16 |

*Source:* CBS, *Automatiseringsstatistieken particuliere sector*, 1986, Staatsuitgeverij, Den Haag, 1988.

processes and products as well as in relation to the development of new activities.

Information technology (IT), also defined as the merger of computer and (tele-)communication technology, is considered a new techno-economic paradigm. Its introduction is able to restructure the whole economy and gives rise to new economic activities. The implementation of the new IT-paradigm not only requires investments in IT equipment but also includes changes in institutions and social and organizational modifications. Examples of former techno-economic paradigms are the steam engine, electric power and chemicals. Although each of these paradigms did penetrate the whole economy, it can be stated that their major applications could be found in agriculture, manufacturing and transport. The distinctive character of the new IT-paradigm is to be found in its penetrating effects into the services sector.

In relation to IT, it is useful to distinguish between administrative applications, on the one hand, and production applications, on the other. Applications in the field of production consist of Computer Aided Manufacturing (CAM) systems such as robots and numerical control equipment. Administrative applications include Computer Aided Design (CAD), Computer Aided Planning (CAP) and different kinds of computers and terminals for purposes of administration and management. While the applications in the field of production are predominantly located in the manufacturing sector, the administrative techniques are applied throughout the economy because their first locus is the office. The variety in penetration rates of administrative IT applications between different sectors in the Dutch economy is illustrated in Table 1. Although the indicators can be criticized in several ways, it is beyond debate that sectors such as banking, insurance and business services make use of information technologies at a relatively high rate.

This relatively high rate of IT application in banking in particular may be explained by different factors. First, the banks' production process is highly suitable for applications of IT, especially in relation to the storage of data (for instance, names, addresses, account numbers and funds held by account holders) and the processing of modifications in data as a consequence of payment transfers, withdrawals or deposits. Second, banks differ from many other sectors in the services sector and in manufacturing as well because they operate on a large scale with a branch network throughout the country. This characteristic made it possible in the first stage of IT to introduce big mainframe computers at central offices which were also used in relation to the activities of the branches. Third, the banks had relatively large funds at their disposal and operated in a growing and profitable market, making it possible to invest in new technologies including software specialists, programmers and educational programmes. Fourth, the dynamics of technological change resulted in the development of micro and personal computers, which, in turn, were suitable for use in local branches. The accumulated knowledge and experience at central offices facilitated a fast introduction and implementation of new technologies at the local level. Consequently, other sectors missing the experiences with mainframe computers at central levels and consisting of individually operating firms are lagging behind.

The early adoption of both mainframe and microcomputers make banking an interesting sector for the analysis of diffusion of information technology. However, before going into the characteristics of a model in order to analyse such a process, it is worth paying some attention to the characteristics of the banking industry itself and its trajectory of information technology.[4] As noted above, the characteristics of the banks' production process made the sector suitable for the application of mainframe technologies. However, in the first phase of computer technology in the 1950s and early 1960s, this suitability was restricted mainly to the field of the payment system. In this respect, it is generally accepted to distinguish between countries that might be characterized as having a 'cheques' payment system (such as the United States and the United Kingdom) and others having a 'giro' system (such as Scandinavian countries and West Germany). It should be understood that the institutional environment influenced the emergence of specific technological trajectories considerably, putting the emphasis on mainframe technologies and automated clearing houses in the case of 'giro' countries and stimulating techniques for the processing of cheques and different modes of 'self-banking' in the case of 'cheques' countries. The payment system in the Netherlands is identified as a 'giro' system.[5]

In his chapter in this volume, de Wit distinguishes between the phase of central automation (1960–80) and the phase of decentralized automation (from 1980 onwards). The first stage was restricted to only a few 'production centres' of commercial banks and at clearing centres; diffusion of new techniques over a larger population was not an issue. During the second phase of decentralized automation, however, we notice differences in the time of adoption of new techniques among branches: in other words, the process of diffusion of information technology starts to become an interesting phenomenon to analyse. Moreover, the effects on employment will no longer be restricted to a small group of production workers at production centres and clearing centres but will apply to the majority of banking employees, about two-thirds of whom are working at local banks' offices throughout the country.

The diffusion model to be developed in the next section will not be of the conventional type used to analyse the spread of just one new technique. Instead, it will be shaped in order to study the diffusion process in a dynamic technological environment, i.e. in an environment in which technological development makes it possible that techniques are replaced by newer generations. We distinguish between four different stages of technological development at the local branch level in banking: (1) conventional 'optical character recognition' equipment; (2) 'back office' terminals; (3) 'front office' terminals; and (4) cash dispensers and automated teller machines.[6] Up to the beginning of the 1970s local branches in The Netherlands were using only mechanical bookkeeping machines. During the 1970s bookkeeping machines came into use which produced output eligible for optical reading at central offices. At the same time, these machines were equipped with controlling functions implying a reduction in errors and 'cleaner' input into the mainframes. At the end of the 1970s and the beginning of the 1980s branches started to adopt minicomputer systems and 'back office' terminals. These terminals were mainly used to process the handwritten payment transfers by account holders. From the mid-1980s, front office terminals were installed in order to automate the processing of withdrawals and deposits by clients at the counter. Finally, cash dispensers were installed in the second half of the 1980s. Before turning to the analysis of the diffusion of these different techniques at the level of local banking, we will first elaborate on the dynamics of diffusion and introduce the model.

## 3. The diffusion of innovations

Some qualitative observations concerning the diffusion of IT have been described in the previous section. In the following sections we will try to embed these observations in a general theoretical framework and express some of its elements in a model.

Dynamics in an economic system originate when a microeconomic impulse induces a macro-scale process of adjustment. Innovation diffusion is a type of economic dynamics, an adjustment of the economy to the opening up of new technological opportunities. It could be characterized as the spread in time of an innovation through certain communication channels among the members of a social system.[7] This means that there are four elements involved in the process of diffusion: an innovation; channels of communication; a social system; and time.[8] Time pertains to our explanandum, to the speed of innovation diffusion. The other elements are the explanatory variables. We shall consider them one by one.

### *Innovation*

The concept of innovation covers items of the most diverse kind. An innovation can involve a process or a product, a technical or an organizational change, an incremental improvement or a radical breakthrough. To clarify the process of innovation diffusion, we draw attention to one dimension in which innovations differ, the location of the innovation. The innovation can be located (1) in knowledge or in a source of information; (2) in a commodity, either for production or consumption; (3) in labour employed; or (4) in an organizational structure.

Some innovations are located in merely one of the above categories. A chemical process may be contained in a formula and be a matter of knowledge only. However, most innovations are located in more than one place. For example, 'front office automation' is a process innovation in banking. It is simultaneously located in journals and manuals, in computer hardware and software, in skills of the employees and in an organizational formula. However, often the 'nucleus' of the innovation is located in one or two of these fields.

The location of an innovation determines some characteristics of the diffusion process. Diffusion implies reproduction and transfer of an innovation. The extent to which an innovation is reproducible and transferrable depends on its type of location. So, referring to the above categories, we can remark the following:

(1) Information and knowledge are usually cheap to reproduce and to transfer, unless they are protected.
(2) New commodities, consumer and capital goods, are reproduced in a traditional production process and transferred by selling. Production capacity limits the availability of an innovation.
(3) The reproduction and transfer of skills is a matter of teaching and learning. Learning capacity limits the adoption speed of an innovation.
(4) The reproduction and transfer of an organizational innovation also depends on learning and on factors like the flexibility of job contracts and organizational hierarchies.

*Communication channels*

Innovations can spread through two types of channels: market channels and non-market channels. One precondition for the functioning of a market is the divisibility of the traded commodity. There can only be a relationship between price and quantity of a marketed good if the good is divisible. Another precondition is the existence of protected ownership. Trade presupposes exclusive ownership. So, whether an innovation spreads by means of market channels depends on whether the innovation is divisible and whether property rights to it can be effectuated.

The location of the innovation is decisive for the determination of the character of the diffusion channels. Thus we observe the following:

(1) It is usually hard to enforce exclusive ownership on knowledge or information. Also, if an innovation is mainly contained in information, it is likely to be indivisible. Thus this type of innovations diffuse predominantly through non-market channels.
(2) Property rights on commodities can usually be effectuated. Goods can be produced in certain quantities. Therefore they can be diffused through market channels.

(3) The diffusion of skills is a matter of training. This process is partly regulated through the market for schooling. However, the relevant price is not only the sum transferred in the market, but also the opportunity costs of the time and effort spent by the trainee.
(4) Organizational change is to a small extent a market-geared process. A new organizational formula can be bought from consultants and specialists in this field.

Since innovations are mostly located in more than one of the above categories, several different channels usually operate in parallel and in connection with each other in the diffusion process. Consequently the speed of innovation diffusion can be slowed down by obstructions in any of these parallel channels.[9] Thus, the spread of computer hardware could, for example, be hampered by the slow development and spread of software or operator training courses.

### *Social system and economic structure*

Markets and other channels function in the context of a social system. We distinguish between two aspects of the social and economic context: (1) the market environment; and (2) the institutional environment.

The relationship between innovative activity and *market environment* has been the subject of extensive research. The usual Schumpeterian conclusion is that some degree of monopoly or monopolistic competition is a necessary condition for dynamic efficiency.[10] The causal ties between market environment and the characteristics of the diffusion process have been less focused upon in research. What can be said about this relationship between environment and diffusion? Three remarks could be relevant.

First of all, the diffusion process is conditioned by a number of parallel markets, as has been argued above. For instance, to explain the diffusion speed of an innovation in process technology, it is necessary not only to take account of the market for capital goods but also to consider the labour and training market.

Second, it can be noted that, as for any market, the supply curve of a market for a process innovation depends on costs and on market form. It is important to point out, however, that the demand schedule in a market for a process innovation is endogenous too. Demand for a process innovation is not derived from subjective preferences, like demand for consumer goods, but from profit-maximizing behaviour on the part of firms producing with the innovation. Their decisions and the restrictions they face depend on their position in their product markets. Thus, the diffusion of process innovations depends also on conditions on the market for the final product that is produced with this innovation.

Third, there is a relationship between the expected rate of innovation and improvement and the diffusion speed. As the market for the innovation is competitive and incremental improvements are expected, potential buyers of the innovation might postpone their purchases and thus slow down the diffusion speed. For instance, demand for today's computers depends on expectations with regard to the development of ever more powerful machines in the future.

The *institutional environment* influences the diffusion speed through its structuring of both markets and non-market communication. Protection of material as well as immaterial property is a matter of institutional arrangements, allowing markets to function. Also, the transfer of information is largely determined by variables with an institutional character. One can think of the access to media and to education, the provision of laboratories, libraries and schools, and the development of postal and telecommunications facilities. To a large extent, this infrastructure has the character of a public good, provided by the government.[11]

### *Markets for innovations and distortions*

To a large degree innovation diffusion is a market process. If we investigate the characteristics of the market for an innovation, the usual instruments of economic analysis enable us to explain the diffusion process. But innovations diffuse partly, to the extent that they take the form of information or knowledge, through non-market channels. The fact that part of the innovation is transferred outside the market distorts the market process. Let us now see why this is the case.

A market functions efficiently under a number of preconditions, one being full and free information about the commodity being traded to all parties involved. Obviously, if this information is part of the traded commodity, this must interfere with the market process. Information available to the traders in the market is (1) incomplete; (2) expanding over time; (3) asymmetric.

First of all, there is considerable uncertainty about both the technical and the economic characteristics of innovations. How does the innovation perform and how profitable is its use? This uncertainty hampers the estimation of the value of the innovation and thus the determination of the demand curve.

Secondly, experience with the innovation builds up in the course of time. This decreases the risk involved in determining the value of the innovation and introducing it.

Thirdly, the supplier of the innovation is much better informed than the potential buyer. Usually the supplier is not willing to reveal all information about the innovation to the market, since this would eliminate the reason to pay for this information and thus destroy the market for the innovation.

From the facts that an efficient market process requires perfect information and that there is no perfect information on markets for innovations, it can be concluded that markets for innovations must be distorted. The distortions are visible in two respects. Firstly, due to lack of information and misjudgements of value, it is likely that there will be trade against non-equilibrium prices. Secondly, because information has a value in the market and is not available without effort, time and resources are spent to collect it. Since collection of information is costly and takes time, market adjustments and thus diffusion are slow.

Summarizing the argument, we can say that the diffusion of an innovation is a process of transfer of some combination of knowledge, goods, skills and arrangements, through the appropriate parallel channels. These channels, to some extent, operate like markets. Their functioning can be distorted by the transfer of immaterial parts of the innovation, information, through non-market channels.

## 4. A model of innovation diffusion

To describe the diffusion of an innovation as a market process it would suffice to identify a demand and supply schedule at any moment in time. The consecutive intersections of these two schedules would give the derivative of the diffusion curve of the innovation. Since the operation of the market for an innovation is distorted by the specific role and the non-completeness of available information, it is (1) impossible to identify the demand curve; (2) likely that the supply curve cannot be derived from increasing marginal production costs; (3) unlikely that the actual price and volume traded are an equilibrium price and volume; and (4) impossible to take account of the gathering and transfer of information as part of the diffusion process.

Therefore, rather than treat diffusion as a market process, we will emphasize in our model the relationship between, on the one hand, the changes in availability of information to the demand side of the market and the time it takes to acquire new knowledge and skills, and, on the other, the introduction of the innovation. Before describing the model, we give a taxonomy of the information, relevant in the process of introducing new production technology.

### *Information as a decisive factor in the diffusion process*

Two stages in the adoption of a process innovation can be distinguished. First of all, there is the decision to invest in some new technique. Secondly, there is the process of learning to operate the innovation. Complex information processing is common to both stages. Thus, the speed of information gathering and processing plays a key role in the determination of the speed of technological progress.

In the first stage of technical change, decision-making, information about investments in alternative techniques has to be gathered, in order to estimate the costs and benefits of switching to a new production technique. In the second stage, the operating procedures of new techniques have to be learned and new skills have to be mastered (see Scheme 1). Apart from technical information, decision making is also based on information on a number of economic variables (see Scheme 2). Whereas the main source of the technical information is usually the supplier of the innovation, the relevant economic information has to come from a large number of sources. Before, during and after introduction of a new technique, skills have to be developed to ensure efficient operation of the technology.

Economic information can be of three types (see Scheme 2). First, we distinguish global information: general facts and knowledge about the sector structure and its working conditions. Then there is firm-specific external information, coming from the environment, its impact depending on the

**Scheme 1** Categories of information

| | Information needed for decision-making: *knowledge* | Information needed for operating: *knowledge and skills* |
|---|---|---|
| Economic information | Global<br>on competitors<br>internal | Production costs |
| Technical information | Performance<br>flexibility | Operating procedures<br>routines |

**Scheme 2** Economic information: examples

| | |
|---|---|
| Global | Product-, inputs- and labour-market development; technological trends; industry structure and macroeconomic development; government policy. |
| On competitors | Competitors marketing strategies (product differentiation, specialization, pricing, geographical spread, promotion) firm structure and production techniques. |
| Internal | Organizational and financial structure; technical features; market position; pricing strategy and costs; firm-specific knowledge and values. |

position of the firm in relation to its competitors, and finally there is firm-specific internal information, depending on the conditions within the firm itself.

In our model we have tried to include some variables that reflect the above two categories of technical and economic information. The technical information is assumed to be exhibited in input coefficients characterizing a technique. With these we calculate a measure for the technical differences between techniques. The larger the disparity between the old and the new technique, the more difficult it is to evaluate and to operate the new technique, the slower its diffusion. The relevant economic information is supposed to be expressed primarily in factor costs and in production volumes of the new relative to the old techniques. First of all, the larger the efficiency gain of the introduction of a new technique, the faster its diffusion will be. Secondly, the extent to which a technique is presently used in an industry conveys information on its expected viability and profitability. The rest of this section will be devoted to an elaboration of these ideas in a model.

The *production volumes* in the present period and the distribution of production over techniques are the direct reflection of the outcome of strategic behaviour in the past. Thus there is a lot of information implicit in production volume figures. Behaviour of the system in the next period is approximated by relating the actual values of production volumes to their potential values. The ratio of the two is a determinant of the speed of change.

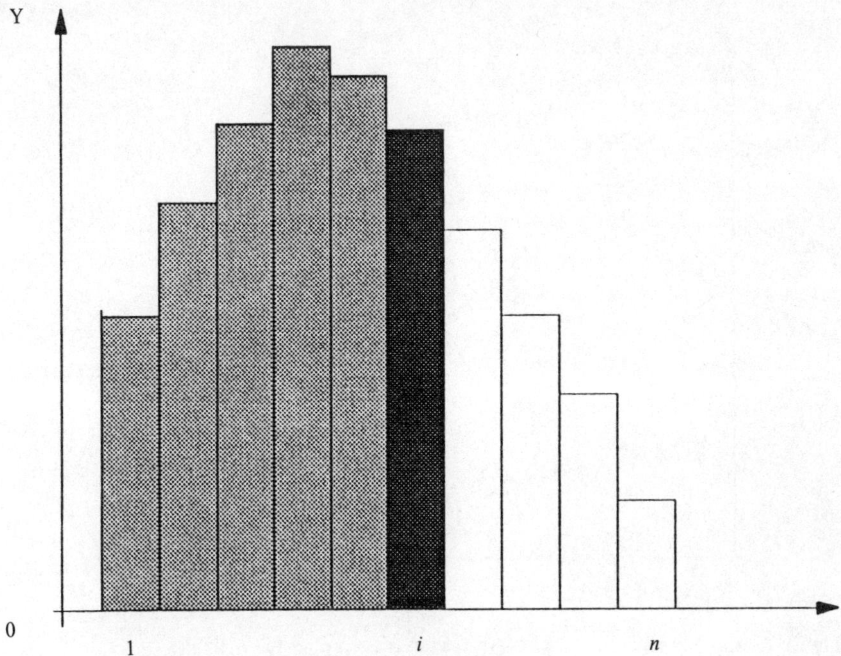


**Figure 1** Production volumes by techniques 1 to *n*

Suppose techniques can be ranked according to increasing efficiency from 1 to *n*. Figure 1 pictures production volumes corresponding to the available techniques. The actual amount produced by means of technique *i* is indicated by the dark-grey column. Under conditions of a constant total production in time, and assuming an efficient use of techniques, the potential maximum amount that can be produced with technique *i* is this column plus the amount produced less efficiently at present, represented by the light-grey columns. This corresponds to the total production volumes of techniques 1 to *i*. The ratio between these actual and potential volumes is used in the expression we call the *competitive force* of technique *i*. Similarly we define the *competitive pressure* on technique *i*, using the ratio of the actual production volume and the maximum effectively competing volume in terms of production costs. The latter is represented by the dark-grey plus open columns in Figure 1 and corresponds to the total production volumes of techniques *i* to *n*.

We characterize production techniques by means of vectors of input coefficients. This enables us to define a distance function in the vector space of techniques. This so called *technical distance* between two techniques is assumed to represent the extent to which these techniques resemble each other. When a firm scraps an old technique and adopts a new one, this distance stands for the technical information barrier that has to be overcome. The concept tries to embody a host of information of varying character, concerning adjustment problems, friction costs, learning efforts and attitude towards risk.

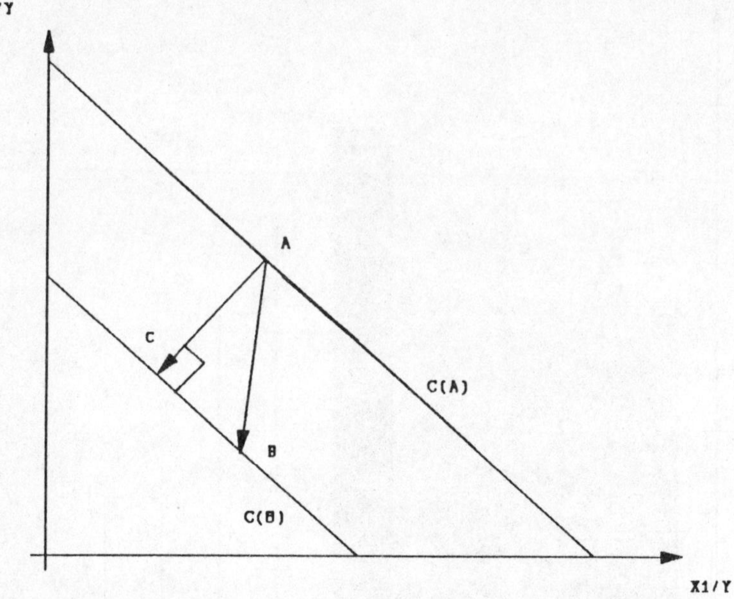


**Figure 2** Costs and benefits when switching from technique A to B

In the model, technical distance is balanced with the *efficiency disparity* variable, which captures the gains from switching techniques. The gains are expressed by computing the cost difference of producing one unit of product by the old and the new technique.

If a technique uses only two factors of production, $x_1$ and $x_2$, the situation can be pictured in a diagram like Figure 2. Suppose a firm operates a technique at A. A new technique is available at B. The lines C(A) and C(B) depend on factor prices and connect all input combinations with equal costs. If a firm switches from the old to the new technique, it has to cross barrier A–B. The benefits of this move are proportional to the distance between the cost lines, distance A–C. The ratio of benefits A–C and costs A–B will determine the extent to which firms are prepared to switch. When factor prices change, this affects the numerator A–C but not the denominator A–B in this ratio.

### *The diffusion model*

The elements introduced above are now combined in a diffusion model. The dependent variable in the model is the change of production volume made by using a technique *i*. Most diffusion models take the number of firms using a particular technique as the dependent variable. Contrary to this practice, we consider production volumes, thereby taking into account that not all firms are equal in size. This method thus gives a more accurate indication of the actual diffusion of a production technique. Also, usually diffusion models only take account of two technical alternatives, the old and the new technique. Our model allows for a large number of techniques to be operated at the same time.

Given the inputs, a range of available techniques and the development of demand, the model can describe the changes in the use of a production technique in the production of a banking product. This change is the result of two components. The first element is the increase in the use of a technique, due to the fact that the technique is adopted by firms that formerly operated less-efficient techniques. The second component is the decrease, caused by the fact that former users of the technique scrap it and switch to a better one. By definition:

$$\text{Change} = \text{Increase} - \text{Decrease}$$

Firms that decide on switching techniques do this by balancing expected benefits against estimated costs. In appraising benefits and costs, they take into account the use of production techniques by competitors as a source of information on their market position and relative efficiency. What is important to a firm, when it considers the virtues of a new technique, is:

(1) the riskiness of adopting a new technique;
(2) the extent of its present use, relative to its potential use: competitive force;
(3) the efficiency benefits;
(4) the technical barriers to be overcome when the firm adopts the new technique.

(ad 1) The risk of adopting a technique *i* is assumed to be due to a lack of information and experience and is assumed to be varying with the extent of its use in the last period : $Y_{it-1}$, where *t* is the time index.

(ad 2) The extent of its present use, relative to its potential use is the ratio of

$Y^{i}_{t-1}$ and $\sum_{k=1}^{i} Y^{k}_{t-1}$, where summation is over the less efficient

techniques. We use this ratio in an expression symbolizing the attractiveness of a technique, called the competitive force – in symbols CF.[12] See also Figure 1.

(ad 3) The efficiency benefits — in symbols EB — are supposed to vary with the cost difference per unit of output.[13] This corresponds, in the two dimensional case pictured in Figure 2, to the distance A-C.

(ad 4) The barrier to cross — in symbols TD — is also expressed in terms of input coefficients.[14] This corresponds to the technical distance A-B in Figure 2.

Thus we can conclude that:

$$\text{Increase} = \alpha_1 * Y^{i}_{t-1} * (EB / TD) * CF$$

where the parameter $\alpha_1$ is a scale factor with respect to the speed of diffusion resulting from competitive force.

Analogously we specify the decrease in the use of a technique. Whereas the impulse to increase the production volume made by using a technique *i* is related to the use of less efficient techniques, the decrease is related to characteristics and volumes of more efficient techniques:

$$\text{Decrease} = \alpha_2 * Yi_{t-1} * (EB'/TD') * CP$$

Here summation is over the production volumes of better techniques when defining competitive pressure CP. The technical distance TD' and the efficiency discrepancy EB' are between technique *i* and the average better alternative. The parameter $\alpha_2$ is a scale factor with respect to the speed of diffusion resulting from competitive pressure.

Pulling all elements together, the change in the production volume made by using technique *i* in period *t* is described by:

$$Yi_t - Yi_{t-1} = \alpha_1 * Yi_{t-1} * (EB / TD) * CF - \alpha_2 * Yi_{t-1} * (EB'/TD') * CP$$

The model describes a life-cycle path for every distinct technique, depending on the speed of innovation of more efficient alternatives. The rise of a technique is modelled as well as its decline. Both the introduction of new technology and the scrapping of old techniques result from an underlying process of optimizing behaviour under conditions of lack of information.

Given input coefficients per technique, we are able to compute technical distances, as can be seen from the definition of TD above. Given wages and factor prices we can determine the efficiency gap EB between techniques. New techniques are expected to be cheaper than older ones, if not from the outset, then at least after a process of learning by doing and increases in capacity utilization. Production volumes are used to compute competitive force and pressure, CF and CP respectively. This being done, the parameters $\alpha_1$ and $\alpha_2$, influencing the general reaction speed to innovations, can be estimated.[15]

## 5. Estimating the diffusion of process technology in Dutch banking[16]

The model presented in the preceding section has been estimated using data from one large Dutch banking organization that consists of more than 900 relatively independent local banks. These local banks all decide on their own at what time to adopt new technology. Out of this total, a representative sample of 100 local banks was included in the data set.

The model has been used to describe changes between 1979 and 1987 in the technique of handling accounts. This product mainly involves the transfers of payments from one account, business or private, to another and the withdrawal and depositing of cash. For account handling four techniques are to be distinguished during this period: (1) conventional: 'optical character recognition' equipment; (2) back office automation; (3) front office automation; (4) cash dispensers/automatic teller machines.

Both optical character recognition (o.c.r) equipment and back office terminals are used for data input activities (cheque and payment orders). Counter terminals and cash dispensers affect data processing (which is done automatically), but primarily deal with teller transactions. Counter terminals came into use in the early 1980s, and rapid growth started after 1983.

*Data*

Given the inputs, a characterization of the four available techniques and the development of demand, the diffusion model outlined above enables us to describe the diffusion of techniques in handling accounts in the branch offices of our case study bank. Local branch offices appear to have different attitudes towards the introduction of new technology. There are early adopters, cautious imitators, slow responders and conservative laggards. Before elaborating on this, we will first briefly describe the main inputs used in the model, together with the relevant factors of production.

In Scheme 3 the different inputs are presented. We sub-divide labour into five categories and we divide capital into a part that does and a part that does not embody technical change: equipment and buildings. Finally we distinguish two other production inputs: mail which also includes intermediary services of the central office, and other factors such as energy, advertising and cleaning. With respect to the labour requirements, it should be noted that these are not measured by numbers of employees or manhours of an employee, but by 'job content', i.e. by hours of activity of a certain type.

The production factors can briefly be described as follows:

- commercial work: negotiations with clients over loans, selling of insurance and travel services, and administrative work like the conclusion and continuation of contracts and handling claims;
- counter work: all kinds of counter activities, such as dealing with deposits and withdrawals and the opening and closing of accounts;
- data input work: consisting of data input for payment transfers and control (checking balances of accounts);
- administrative work: other administrative activities including accounting, catering services and small repairs;
- managerial work;
- central computer processing and mailing (ccpm): processing of transfers, storing accounts and savings information and mailing at the central office;
- inventory including computer equipment: inventory and computer and communication systems at the local branch office;
- buildings;
- remaining inputs: energy, advertising and cleaning.

**Scheme 3** Factors of production in banking

| *Labour* | *Capital* | *Other* |
|---|---|---|
| Commercial work<br>Counter work<br>Data input work<br>Administrative work<br>Managerial work | Buildings<br>Inventory including computer equipment | Mail<br>Other |

Vectors describing a technique will have nine elements, corresponding to the production factors in Scheme 3. Furthermore, prices of the various capital services, mail and maintenance, as well as wages for five different types of labour, are needed. Finally, we need the production volumes made every year by the distinct techniques.

For the 1973–87 period, cost figures for inputs, employment figures and production volumes of products were collected from the 100 banks. Price indices for inputs stem from local offices (like labour costs per man year) and from other sources (price indices on computer equipment coming from the Central Bureau of Statistics in the Netherlands). Production volumes for accounts are based on the number of payment transfers, teller transactions and the number of payment accounts, both personal and business.[17]

In every local branch office in every time period a technology for payment transfers is identified through detailed data on automation from individual offices. We also know production volumes and inputs per branch office. This makes it possible to determine aggregate production and input coefficients for each technique in use. Production volumes for techniques are computed by aggregating the production volumes of local offices which used the same kind of technology. Input coefficients for accounts are calculated by aggregating the costs of inputs of offices using the same technology and deflating these costs by their price index, and dividing this by their (aggregated) production volume. For the entire estimation period input coefficients are computed for each technique, not only showing differences in the amount of an input per unit of output between techniques, but also within a single technique over time.

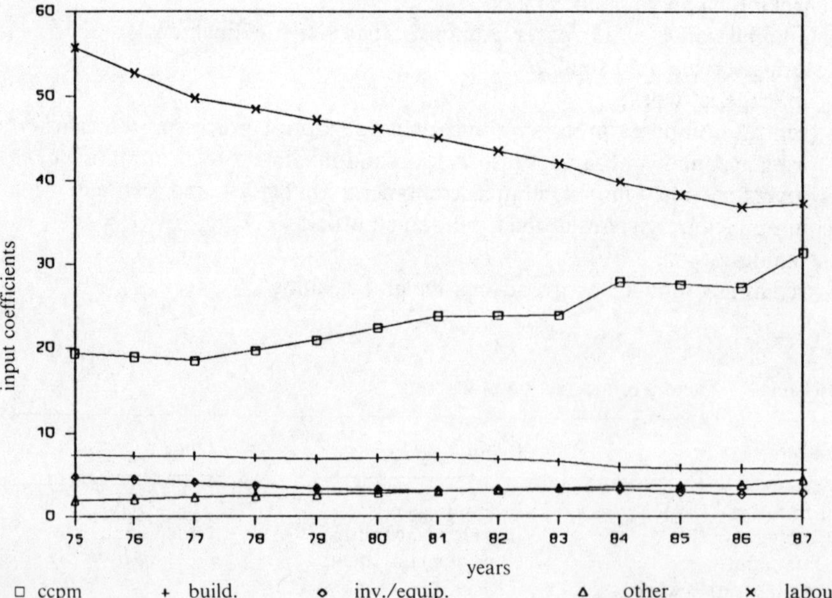


**Figure 3** Input coefficients accounts

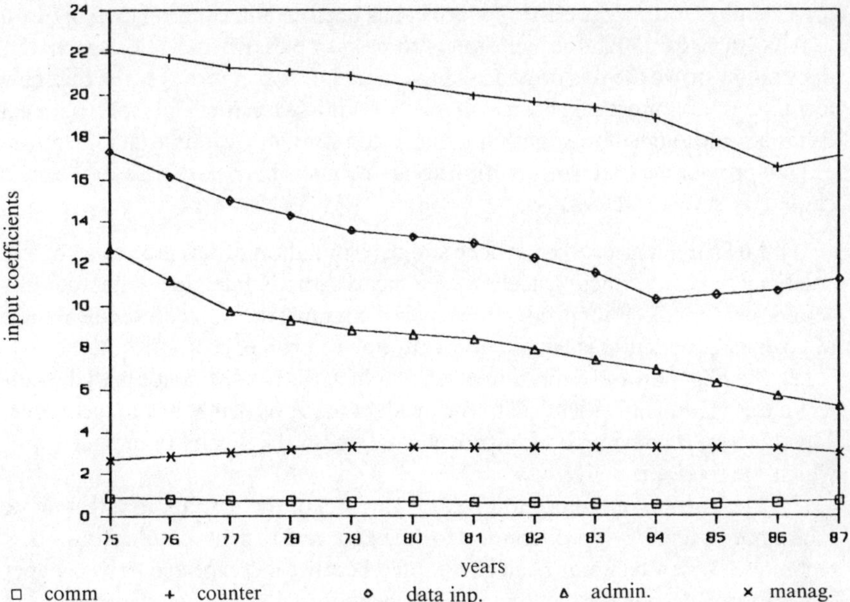


**Figure 4** Input coefficients accounts, labour

The aggregate input coefficients are presented in Figures 3 and 4, for the years 1975–87.[18] Accounts experienced substantial labour savings. Automation is reflected (directly) through input coefficients for computer equipment and inventory. Figure 3 clearly shows how input coefficients for computer equipment increased.

Looking at the decomposition of labour input per unit of output for accounts in Figure 4, we see that technological change decreased the input of counter work, data input work and administrative work. Although labour input per unit of output decreased, employment increased in this period due to the growth of production volumes (5.3 per cent in the whole 1975–87 period).

Comparing disaggregated input coefficients for different techniques of account handling shows that input coefficients for computer equipment were higher for every consecutive technique. On average, each new technique leads to less labour input, mainly as a result of less data input work and less administrative work. Labour savings for counter work do not appear in the data. Implementation of counter terminals does not lead to large labour savings, at least initially. Also in the case of cash dispensers, input coefficients do not reflect the labour savings on counter work.

### *Estimation results*

Measures for competitive forces and competitive pressures of techniques have been computed using production volumes of techniques in the previous period. Efficiency gaps and technical distances between techniques have been calculated for several years and the diffusion model has been estimated using these figures.

It has turned out that the inclusion of these variables, EB and TD respectively, does not improve the performance of the model. It sometimes appears that new techniques are more expensive than old ones. Measurements of technical distances, although of the right sign, fluctuated strongly over time, and therefore did not provide accurate information about technical barriers. These facts can be caused by several factors.

(1) Measurement problems: accurate determination of a vector of technical coefficients per production technique is nearly impossible, due to factors like economies of scale, joint production, special circumstances; accurate calculation of costs of production is equally difficult, due to problems of attribution.

(2) Aggregation and definition of the product: the product made with the new technique is not really identical to the product made by using the old technique. Mostly there is some type of improvement, e.g. in the flexibility or speed with which the service is delivered.

(3) The current difference in production costs of the new and old technique might not be a good approximation for the expected benefits of adopting a new technique. A new technique can be adopted because it is expected to be cheaper in the future, despite the fact that it might be more expensive in the introduction phase. Due to learning by doing and economies of scale the price of a new technique may fall considerably after first introduction.

(4) The efficiency gain may not be one of the main arguments for adopting a new technique. The firm is merely afraid to fall behind in technology or mainly motivated to keep up with its competitors.

(5) At the moment of introduction the new technique does not operate at full

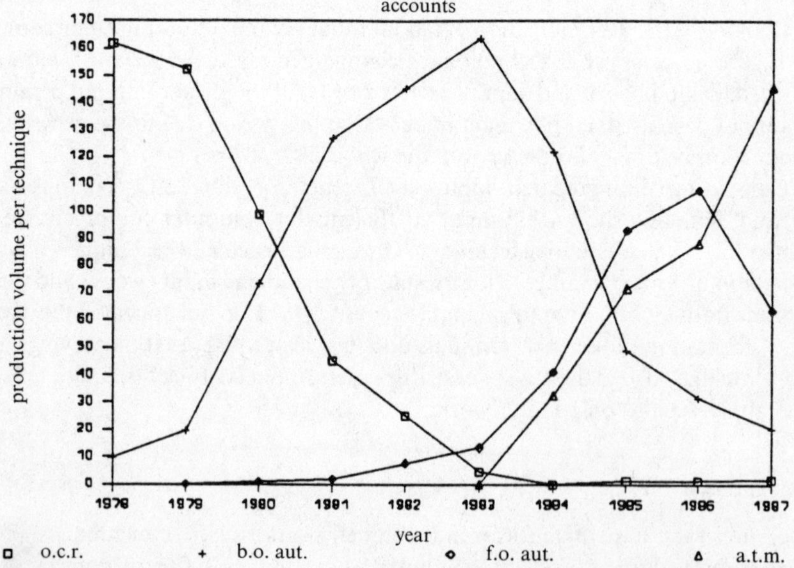


**Figure 5** Diffusion of techniques

capacity. Investments in a new technique often lead to excess capacity, which in the short term disproportionately raises the costs of the new technique.

The diffusion process of technology was best explained by competitive forces and competitive pressure as the only variables. Model results for this simple version of the diffusion model will be given. The estimations are produced by minimizing least squares (under the restriction of $\alpha_1$ and $\alpha_2$ being the same for all tecniques, but not that $\alpha_1$ and $\alpha_2$ are equal). Market expansion is reflected by the estimates for $\alpha_1$ and $\alpha_2$, 1.45 and 0.74 respectively, $\alpha_1$ being almost twice as high as $\alpha_2$.[19] In Figure 5 real production volumes of techniques for accounts in the 1978–1987 period are presented. In Figure 6 the production volumes for techniques as explained by the diffusion model are given.

Figure 5 shows how conventional ocr equipment is substituted by back office computer equipment. Back office terminals, first adopted in 1978, experienced a fast growth within five years. In 1983 nearly all banks possessed back office computer systems. In the years following 1983, front office terminals and cash dispensers were implemented. Like back office terminals, front office terminals grew rapidly in a five-year period. Cash dispensers in local branch offices experienced an even faster growth. Most likely this fast growth will continue in the coming years.

Changes in the choice of technology, among other factors, lead to changes in factor demand. Since each technology is characterized by input coefficients, reflecting the amount of input per unit of output, the consequences of technology diffusion on factor demands can be computed. However, in order to calculate overall changes in factor demands of the banking industry, the above estimation exercise has to be performed for every banking product. We have done this in

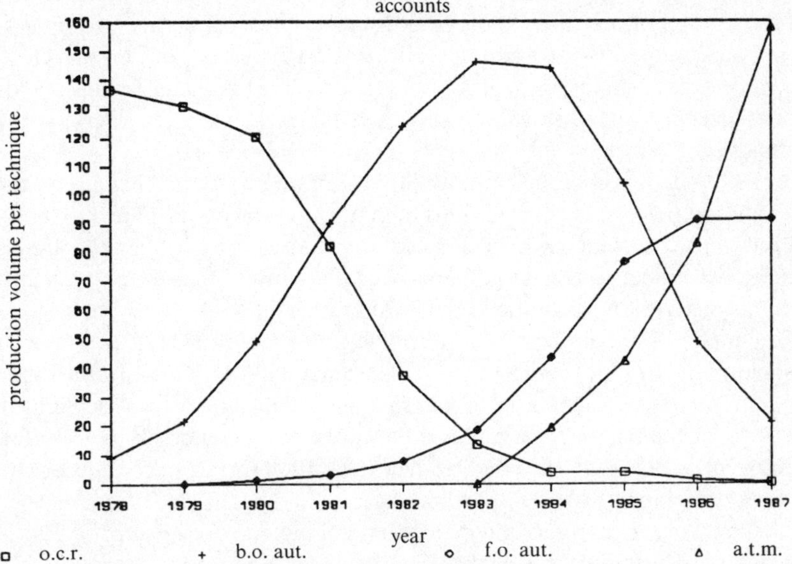


**Figure 6** Diffusion of techniques, model results

Diederen *et al.* (1989), where we also looked at the implications of the diffusion of these techniques for the occupational structure and schooling requirements in banking. A further discussion of the results of that study, however, lies outside the scope of this chapter.

## 6. Summary and concluding remarks

In this chapter we have dealt with the diffusion of applications of information technology in Dutch banking and presented a diffusion model to explain several of the developments in more detail. We discussed the nature of information technology (IT) and its penetration in various branches of the economy. Since banking is a highly IT-intensive sector, we elaborate on this as an example. For this reason, technological development in banking in general and the successive techniques employed in local branch offices in the last decade to handle accounts are briefly described. The latter techniques are interesting, since their diffusion over local banks can clearly be observed. It is to this process that our diffusion model is applied.

The diffusion model is based on the notion that introduction of new technology requires accumulation of information and opportunity to learn. Moreover, the social and economic environment plays an important role in the acceptance of new technologies. Therefore the availability of a more profitable technique does not imply that it will be used immediately: different techniques will be employed at each moment to produce a certain product. This is a consequence of phenomena such as: the limited capacity of individuals and organizations to process multi-type information; the balancing of current expenditure against uncertain future benefits; and the dependence of firm decisions on perceived reference groups. These phenomena are captured in the model by the notions of technological distance, efficiency barrier, competitive power and competitive pressure. The notions lead to two expressions that closely resemble logistic curves, one for the growth in the use of each technique and one for its decline. Together they result in a life-cycle path for every distinct technique.

In the model a technique employed for a certain product is characterized by its input coefficients, i.e. the amounts of inputs necessary to produce one unit of output. Since this requires very detailed information, we confined the application of the model to one large bank over the period 1979–87, for which we obtained detailed data on about 100 local branch offices.

The model results indicate that technological change has had an important impact on factor inputs with respect to some banking activities. In the case of accounts there were substantial labour savings on data input work (including the checking of the balances), administrative work and counter work. Computer equipment in the back office led to savings on the processing of transfers that involve data input work and the checking of the balances of accounts, and further, personal computers and main frame computers, storing information, led to savings on administrative work.

The development of the input coefficients of a single technique over time reflects the notion that technological change involves learning and adjustments. Benefits that are related to the new technologies are not captured immediately. Employees must be retrained and get used to new technology: it takes time to learn to work more productively. Therefore, a necessary extension of the model is to allow for the fact that techniques diffuse which are temporarily inefficient, at least in their initial stages of introduction. In this context the relationship between technological change and new products is also important. It can be observed that technical change has induced changes in commercial policy in banks. There has been a move from a product-oriented approach to an approach called 'integrated client management', where customers are served the whole range of banking products by a single 'personal banker'.

Another important extension of the model would be to be more explicit on what elements in the evolution of specific firms are favourable to the introduction of new techniques. The skill structure of the workforce is an obvious candidate. Finally, the change in input coefficients over time of existing techniques should also be explained. This change can be considerable. These possible elaborations point at limitations of the model in its present form. However, the possibility of identifying these limitations illustrates that a model is a very useful way to analyse causal relationships. Moreover, in spite of these deficiencies, the model fits the data quite well. And one should realize that by using the model some causal relationships between variables could be firmly established.

In a certain way it is a challenge — and maybe even more than that — to look at the possible policy implications that can be inferred from the above analysis. One should be careful, however, for various reasons. First, neither all policy ends nor all instruments are distinguished separately, if at all, in the model. With respect to the ends, the policy might aim at enhancing diffusion, but that will always be an intermediate step in order to reach a further goal which lies outside the scope of the above analysis — for example, full employment or export promotion. Moreover different means of stimulating diffusion might affect the attainment of this further goal in different ways. For instance, subsidizing the adoption of new technologies and an active policy of using new technologies in government agencies will obviously have different employment effects compared to a policy aiming at a better dissemination of information on new technologies, e.g. through information centres, and providing better facilities for (re)schooling.[20] This example also illustrates the other reasons to be careful when giving policy recommendations. For by its very nature the analysis is partial, focusing on the diffusion process. This implies that even if it is possible to evaluate the impact of a certain measure on the diffusion process, and maybe even on the ultimate goal, the side effects of applying this measure with respect to the achievement of other goals mostly fall outside the scope of the present analysis. In that respect the recommendations are always of a highly partial nature.

Now these are rather general remarks which apply to most policy advice. Essentially the problem is that there is not an all-comprehensive model of the economy in which all variables are identified and all possible interactions are captured in a dynamic setting. However, with respect to the diffusion of technologies this problem is particularly severe, since the diffusion process is of a highly dynamic nature and interacts strongly with other processes in the economy. This has been highlighted in the analysis in Section 3. It is therefore also not surprising that literature on diffusion policy is scarce.

None the less, the above analysis can provide some useful insights which should be taken into account when policy proposals are formulated. Underlying these insights is the notion that the diffusion of IT, both with respect to process innovations and to product innovations, is an important phenomenon with, amongst others, far-reaching consequences for the size and quality of employment. These consequences have been illustrated above for the case of banking. This implies that policy measures aiming to control or increase the speed of technological change, and perhaps also influence its direction, should not concentrate on innovation policy alone, as often is the case. Specific measures should also be considered in order to control or speed up the diffusion of these innovations. We will discuss some measures briefly below. On the other hand one should be alert to recognize a possible impact on the diffusion of technology of policy measures directed towards other goals. For instance, a policy of employment creation by systematically keeping wages low in connection with attracting low-grade employment might hamper the diffusion of new technologies and finally be self-defeating in a situation where product quality and technological flexibility are important features of the production process. An example is the poor performance of several British branches of industry compared to German ones.[21]

In Dankbaar *et al.* (1989) several policy measures to stimulate diffusion are elaborated. However, these measures should be seen in the light of an experimental attempt to derive policy implications from the above analysis. The importance of an active information and schooling policy is emphasized. One aspect of the information policy is that information on the introduction of new technologies and its impact should be gathered systematically in an early stage. To this end a special agency should be created or a new division should be added to an existing agency. Another aspect is to stimulate the provision of information in technologically backward sectors both on new technologies and on solutions for potential organizational problems. Information on these topics is relevant anyhow for small firms — to which special attention should be paid. With respect to the extent of potential organizational problems, the technical distance of new technologies, which was introduced in the context of the diffusion model in Section 4, might be an interesting indicator. It seems plausible, that the larger this distance, the more organizational changes will be necessary.

With respect to schooling policy the generic character of IT should be taken into account. In formal schooling this should be reflected by paying more specific attention to the nature of technical systems and the interaction with organizational structures. Thus fields like systems analysis, process control, computer use, management and marketing should be taught on a wider basis. Further application of this process-oriented knowledge in particular sectors should be learned in sector-specific training institutions or by on-the-job training.

It is evident that these measures are not implied in any strict, rigorous way by the above analysis. This illustrates the earlier reservations on formulating policy implications. However, they are consistent with the above analysis and can be interpreted in the light of it. And it is more in that spirit that from our point of view policy advice should be given. That is, the analysis should not primarily be seen as a means to generate proposals but more as a framework of reference in the light of which policy proposals can be interpreted and evaluated. Such a framework is important to highlight the importance of certain variables, to identify relevant interactions with the economic and social environment and to check the consistency with other measures. Of course, in a discussion with policy-makers the above analysis is not a fixed bench-mark. On the contrary it is obvious that such a discussion can also have a stimulating influence on the further development of the analysis presented above.

## Notes

1. This chapter is based on research done within the framework of the CERI Programme of the OECD and on work done for the Ministry of Social Affairs and Employment in The Netherlands. Results of these projects are reported in Diederen *et al.* (1989) and Dankbaar *et al.* (1989). We thank Donald McFetridge for his stimulating comments.
2. See Freeman (1987).
3. See Perez (1983), Schumpeter (1939), Nelson and Winter (1977) and Dosi (1983).
4. A fuller description is given in the paper by De Wit in Chapter 5 of this volume.
5. See De Wit, Chapter 5, pp. 100–102.
6. This is in line with De Wit (Chapter 5, p. 101).
7. Rogers (1983), p. 5.
8. Mahajan and Peterson (1985), p. 7.
9. Diffusion studies, notably the ones using probit models, mostly treat diffusion as if only market channels with known prices are the relevant transmission channels.
10. See e.g. Kamien and Schwartz (1982) or Baldwin and Scott (1987).
11. The development of traditional transport infrastructure provides a good illustration of the influence of the (institutional) environment on the diffusion of an innovation in transport, the car. The build-up of an infrastructure of roads, highways, gas stations and driving schools is closely connected to the development of the car and was a precondition for the massive diffusion of car transportation.
12. $CF = 1 - \dfrac{Y_{i_{t-1}}}{\sum_{k=1}^{i} Y_{k_{t-1}}}$
13. $EB = \sum_{k=1}^{m} w_k(x_{ik} - x_{jk})$, i being the new and j being the average old technique and summation being over all input coefficients $x_k$.
14. $TD = [\sum_{k=1}^{m} (x_{ik} - x_{jk})^2 ]^{1/2}$.
15. Due to disembodied technological change, e.g. as a consequence of learning by doing, techniques can change marginally over time. This process can influence their relative costs and technical distances. Therefore, we measure technical coefficients of one technique at several points in time. We expect the size of the $\alpha_i$s and $\alpha_2$s for different products to reflect the toughness of competition in the distinct product markets. The $\alpha_1$s will be slightly larger than the $\alpha_2$s in expanding markets.
16. This paragraph is derived from Diederen, Kemp, Muysken and De Wit (1990).

17. Resulting production volumes for accounts and savings are composed out of these components by weighting their importance by unit costs: for accounts, the number of payment transfers, teller transactions (deposits and withdrawals), and the actual number of payment accounts (to include the actual establishing of an account) are aggregated according to the ratio of their unit costs, i.e. 1:4:60.
18. The input coefficients, depicted on the vertical axis, give the amount of input per unit of output, as computed by price indices for inputs and composed production volumes for products.
19. It is hard to draw conclusions with respect to the competitiveness of the market (that is reflected by the absolute value of the estimates) because the values for $\alpha_1$ and $\alpha_2$ cannot be compared with the values in other sectors.
20. See Stoneman (1987), Ch. 6 and Ch. 7 for an elaboration. He argues that under certain restrictions information policies should be preferred to subsidization policies when the industry supplying new technologies is competitive, whereas the reverse is the case under monopoly supply (p. 79).
21. See e.g. Steedman and Wagner (1990).

## References

Baldwin, W.L. and J.T. Scott (1987), *Market Structure and Technological Change*, Chur: Harwood Academic Publishers.

Dankbaar, B., Diederen, P.J.M., Hagedoorn, J., Kemp, R.P.M., Muysken, J. and G.R. de Wit (1989), *Diffusie van technologie en de ontwikkeling van de werkgelegenheid: toepassingen voor diensten, bouw en industrie*, Maastricht, September.

Diederen, P.J.M., de Grip, A., Groot, L., Heijke, J., Kemp, R.P.M., Meijers, H., Muysken, J. and G.R. de Wit (1989), *Technological Change, Employment and Skill Formation in Dutch Banking*, Maastricht, April LIB/W/002.

Diederen, P.J.M., Kemp, R.P.M., Muysken, J. and G.R. de Wit (1990), 'Diffusion of process technology in Dutch banking', in *Technological Forecasting and Social Change* (forthcoming).

Dosi, G. (1984), 'Technological paradigms and technological trajectories', in Freeman, C. (ed.), *Long Waves in the World Economy*, London: Pinter.

Freeman, C. (1987), 'Information technology and change in techno-economic paradigm', in Freeman, C. and L. Soete (eds), *Technical Change and Full Employment*, Oxford/ New York: Basil Blackwell.

Kamien, M.I. and N.L. Schwartz (1982), *Market Structure and Innovation*, Cambridge University Press.

Mahajan, V. and R.A. Peterson (1985), *Models for Innovation Diffusion*, Beverly Hills: Sage Publications.

Nelson, R.R. and S.G. Winter (1977), 'In search of a useful theory of innovation', *Research Policy*, pp. 36–76.

Perez, C. (1983), 'Structural change and the assimilation of new technologies in the economic and social system', *Futures*, 15(4), pp. 357–75.

Rogers, E.M. (1983), *Diffusion of Innovations*, New York: Free Press.

Schumpeter, J.A. (1939), *Business Cycles*, New York: McGraw-Hill.

Steedman, H. and K. Wagner (1990), 'Productivity, machinery and skills: clothing manufacturing in Britain and Germany', in Dankbaar, B., Groenewegen, J. and H. Schenk (eds), *Perspectives in Industrial Organization*, Dordrecht: Kluwer Academic Publishers (forthcoming).

Stoneman, P. (1987), *The Economic Analysis of Technology Policy*, Oxford: Clarendon Press.

# 7. Information technology introduction in the big banks: the case of Japan

*Yasunori Baba and Shinji Takai*[1]

## Introduction

The growing competitiveness of Japanese manufacturing has prompted a number of studies which indicate the varied sources of this success. Light has been shed not only on the 'Japanese national system of innovation' (Freeman, 1987, 1988) at national level, but also on managerial and organisational innovations made at enterprise level. Managerial innovations on the shop floor (e.g. the QC circle) and organisational innovations in product development (e.g. overlapping development phases (Imai *et al.*, 1985)) have become virtually common knowledge. However, is there not a 'black box' still left unopened? Given the impressive productivity record of Japanese service sectors (compared, for instance, with the pattern of decline in the US),[2] further questions have to be answered. Do Japanese manufacturing techniques have their counterparts in the country's service sectors, and, if so, how do they contribute to the productivity growth in services?

To begin answering these questions, we shall take as an example the Japanese banking sector, where firms are currently speeding up their drive to introduce information technologies (ITs). Our illustration will explain in historical terms (*1*) how firms initially introduced ITs into their daily operations; and (*2*) how they have gradually translated their preferred methods into a distinct corporate strategy. Then, after making reference to the way IT introduction has proceeded in the US, we will suggest how the new Japanese services techniques have actually contributed to productivity growth and the efficient day-to-day operation of the banking sector. On the basis of this analysis, the chapter will end with some proposed insights for banks preparing for heightened global competition.

## The organic phase of IT introduction, mid-1960s to the 1970s

In the Japanese banking sector, the first phase of on-line IT operation[3] began in the mid-1960s. At this stage, the introduction was somewhat fragmentary, limited to individual operations of parts of the accounting system (i.e. deposit and exchange), and confined to an intra-bank networking range. Not until the mid-1970s did introduction move into a further stage. The second phase on-line[4] enabled the banks to systemise each unit of their accounting systems, and the range of networking was now inter-bank. Consequently, new services like the inter-bank cash dispenser (CD) and automatic teller machine (ATM) appeared on the market and contributed to the banks' marketing. But, while their business was being transformed in this way, how did the banks deal with the managerial problems concomitant with IT introduction? In concrete terms, what kinds of Japanese services techniques actually translated day-to-day routine back-office operation into detailed, workable programmes?

To find out, it is possible to identify the contents of on-site operational procedures partly through computer modelling where there is an essential logic to the procedures, and partly through observation of ongoing operations, and then codify them into an operational manual. System engineering transforms this into workable programmes. Taking the case of preparations for ATM, programmes have to be articulated to fit in with the registering methods of the existing passbook system. Inter-bank ATM will not work unless banks' registering methods are standardised.

However, in contrast to this method, what the Japanese banks actually did (in accordance with the Japanese tradition of bottom-up decision-making, where 'the use of on-the-spot knowledge and rapid problem-solving through learning-by-doing [Aoki, 1986] is encouraged), was to formalise their computer routines on the basis of their accumulated on-the-spot procedures. Collaboration (e.g. communication and information-sharing) between profit centres and IT departments (specialising in system analysis and development) became a crucial part of organisational arrangements. A key role in efficient collaboration was played by link-persons who had become acquainted with both daily operating procedures and system engineering, through having followed a career path where promotion (or rotation) results in an accumulation of skills and knowledge through on-the-job training in a succession of different posts. Even where banks bought software in the market, the same philosophy was applied: the back-office workers modified the software to render it compatible with existing operational procedures. Through processes of incremental modification, the banks were able to provide their specific business operations with a good deal of user-friendly software. Generally speaking, the banks' computerised routines were devised to develop mainly on a bank-specific basis.

## The pragmatic phase of IT introduction from the early 1980s

In contrast to the organic phase, where the traditional business culture of Japan directed IT introductions more or less spontaneously, the early 1980s seem to have witnessed a new phase. After a trial stage of IT introduction, the banks were under strong commercial pressure to utilise ITs fully (e.g. introducing new services like firm banking etc. into an existing infrastructure). Under threat of a rise in back-office work load in connection with IT growth, the banks' common aim became to introduce ITs and then *utilise* them productively and efficiently. The banks rethought their strategies for IT introduction, trying to formalise a practical package for the use of ITs in their daily operations.

### *Organisational drive for adopting user-friendly systems*

Let us firstly summarise how banks built up organisational decision-making

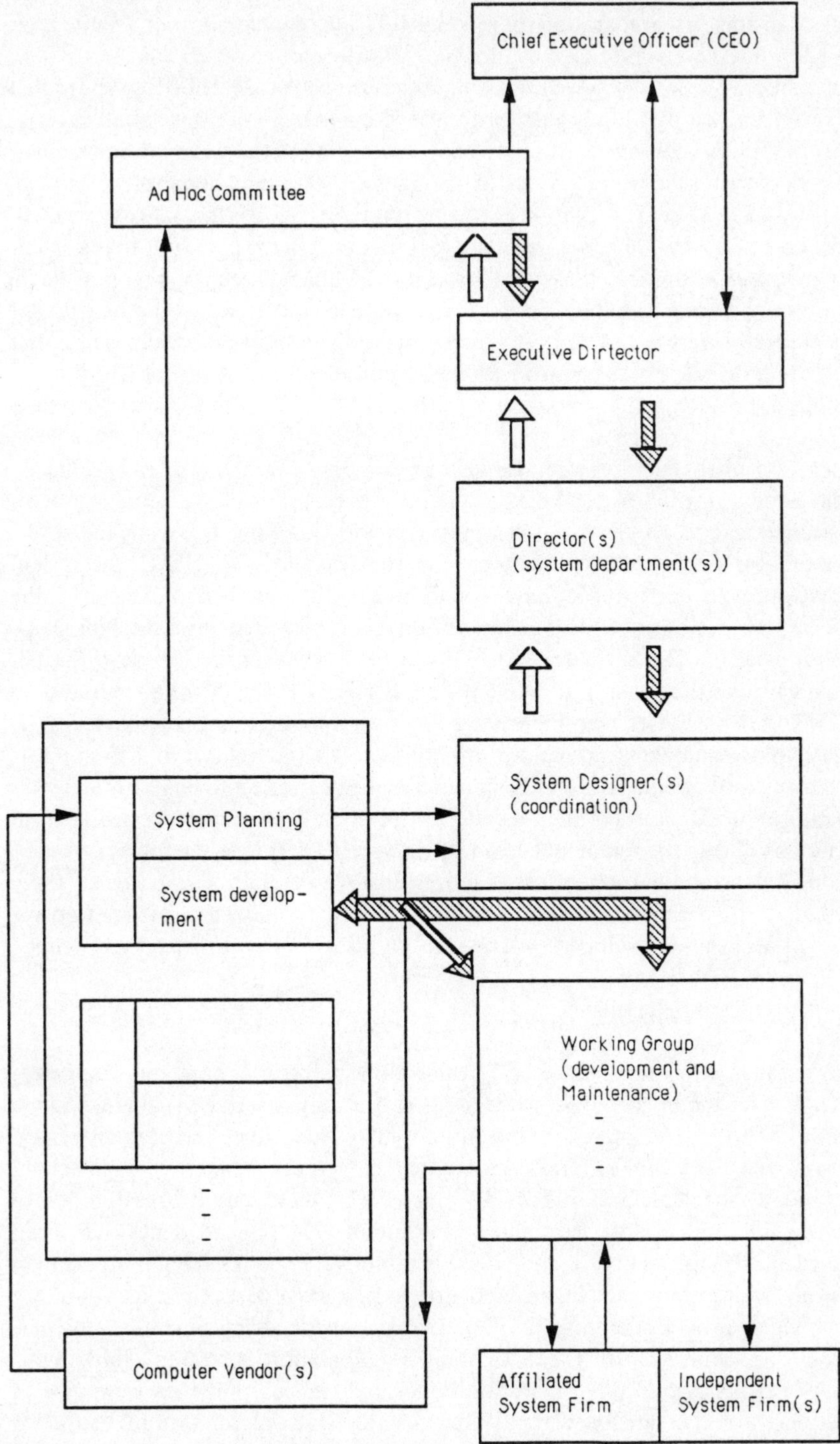


**Figure 1** Decision-making process of IT introduction

mechanisms as a preparation for piecemeal IT introduction or modification (see Figure 1). This was generally done in profit centres which find it pays to introduce IT into a given existing operation. As described, well-trained specialists identify a likely plan for IT introduction from the angle of managerial profitability. These specialists then propose the plan to the system department in terms of system-concept and system-language. The system department, in turn, carries out a system evaluation of the proposal on cost/performance criteria. In the case of a minor change, the board gets involved in the plan only to the extent of giving *ex post facto* approval. However, when banks attempt a large-scale IT introduction (e.g. the third-phase on-line system[5]) a decision at top board level is naturally required. A project team comprising executive directors is often set up to report back to the board with their opinions and judgements. The process tends to be managed in an *ad hoc* fashion, so that it sometimes takes a long time to reach a final decision.

When a firm decision is made, whatever it is, how are the systems specifically developed (or modified)? As far as the intra-bank system development process is concerned, we can list the following characteristics. First, following the model of the job sharing arrangement used in the system design phase, the system development department now collaborates with intra-bank users in the development phase too. This collaboration seems to contribute to meshing shop-floor needs with the systems actually developed. Second, the development process, like that seen in manufacturing, is associated with spiral-overlapping development phases (see Figure 2). The overlapping is actually between the design, development and maintenance phases: when development costs appear to overshoot planned limits, the tendency is for the project to be reconsidered in terms of modification of the original design; when users experience problems in systems during the maintenance phase, their reports trigger the introduction of a new development phase for a more sensitive version. The result of these characteristic methods is to make possible mutual adjustment between operational needs and developed systems, and to bring forth user-friendly systems.

### *An emergence of flexible, intelligent back-office*

In parallel with the advance of IT utilisation, banks overhaul and renovate a whole package of corporate strategies. Let us firstly describe the changes made in human resource management, in particular back-office recruitment policy. From the early 1980s, banks started reducing their reliance on high-school graduates as female back-office staff, and have begun to recruit large numbers of university and college graduates (see Figure 3). This tendency has been particularly conspicuous among highly IT-motivated banks. Thanks to the high turnover rate among the high-school graduate workforce (the average voluntary leaving age was in the range 22–23), the banks were able to introduce the new policy and eliminate the excessive numbers of back-office staff in a fairly light-handed way (see Figure 4). In addition, banks hoped that their new type of recruits would create a new type of back-office. These recruits with far superior qualifications and overall ability, provided with a whole package of training, are expected to compose a body of multipurpose support staff. The new-vintage

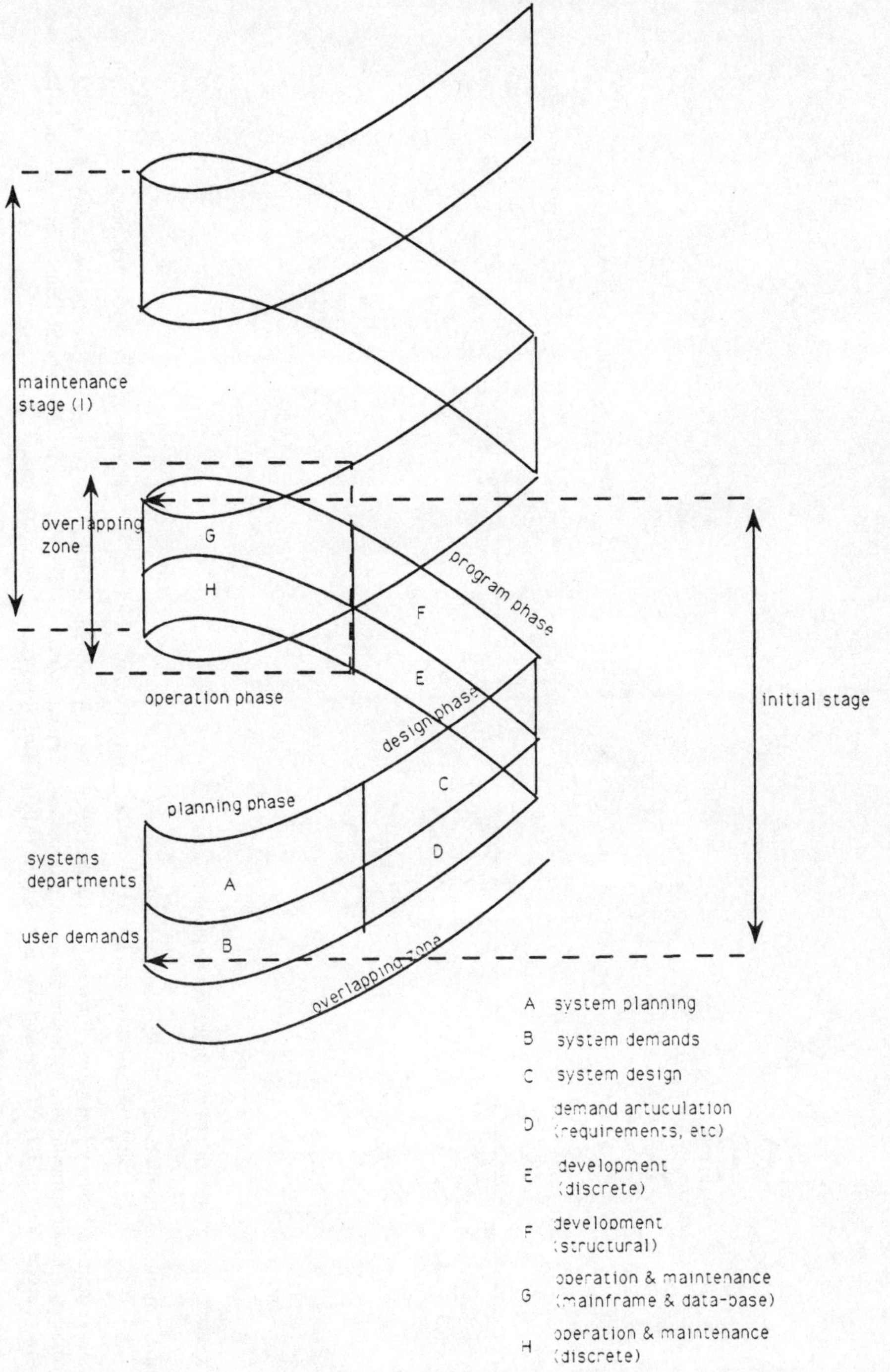


**Figure 2** Spiral overlapping process of software evolution

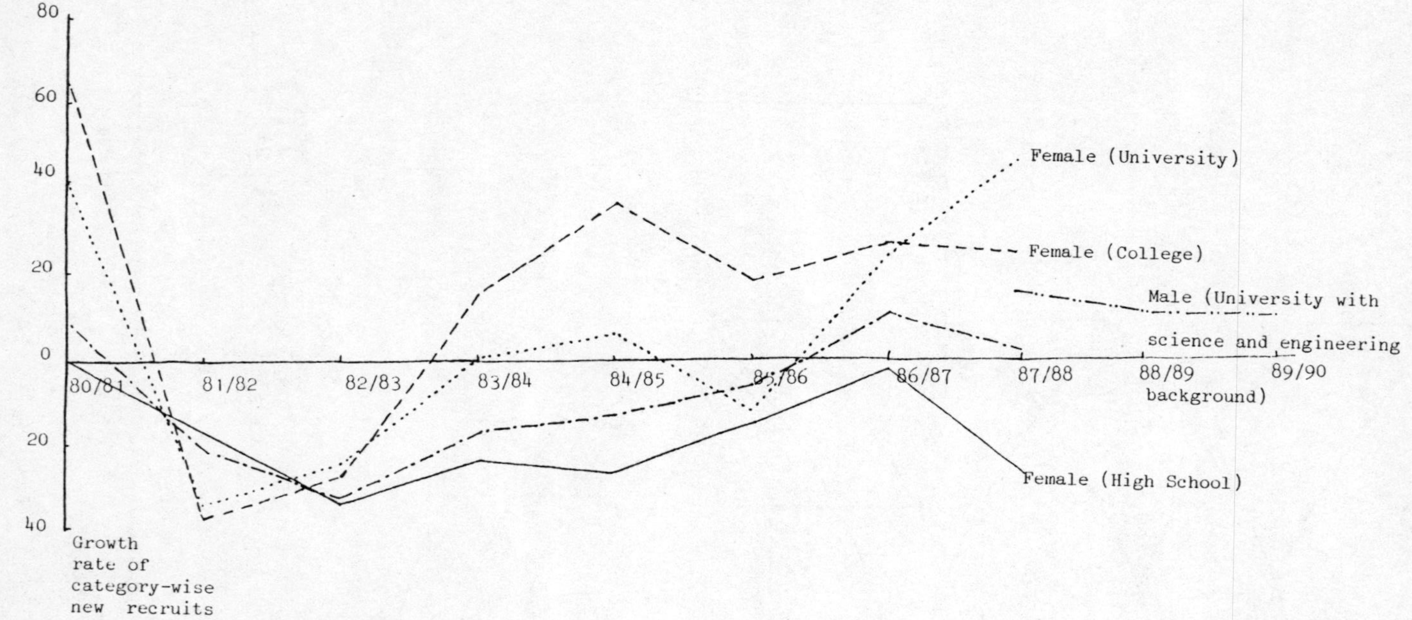


**Figure 3** Changing recruitment policy (male/female) through the 1980s

*Source:* female numbers: using data from *Nikkin Financial Directory* the Japan Financial News Co., Ltd.; male number: using data from *Weekly Toyo Keizai*, 30 September 1989.

*Note:* Female sample includes Japanese big thirteen city banks (DKB, Mitsui, Fuji, Mitsubishi, Kyowa, Sanwa, Sumitomo, Daiwa, Tokai, Takugin, Taiyokobe, Tokyo, Saitama) and male sample includes Japanese seven city banks (DKB, Mitsui, Fuji, Mitsubishi, Sanwa, Sumitomo, Tokai), two long-term credit banks (IBJ, LTCB) and five trust banks (Mitsubishi, Sumitomo, Mitsui, Yasuda, Tokyo).

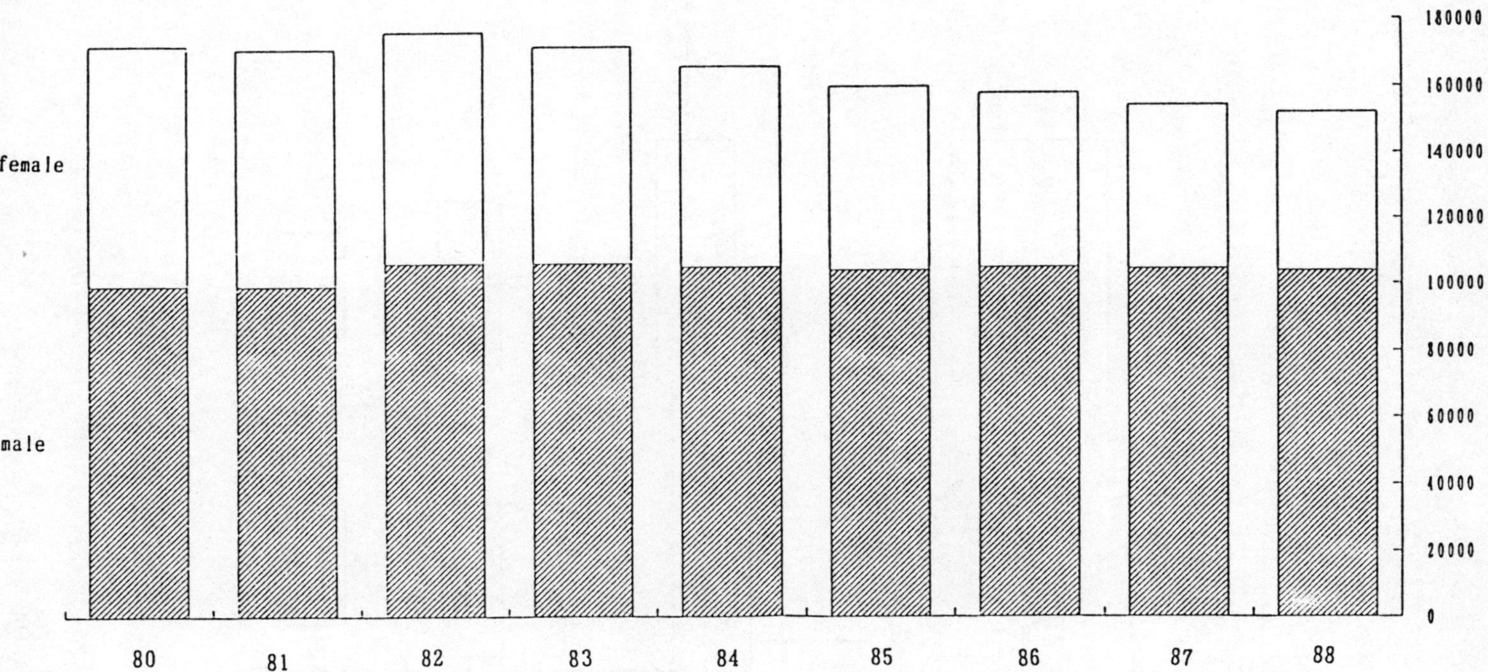


**Figure 4** Changing numbers of bank-officers (male/female) through the 1980s

*Source:* Using data from annual reports.

*Notes:* Sample includes Japanese big thirteen city banks, (DKB, Mitsui, Fuji, Mitsubishi, Kyowa, Sanwa Sumitomo, Daiwa, Tokai, Takugin, Taiyokobe, Tokyo, Saitama)

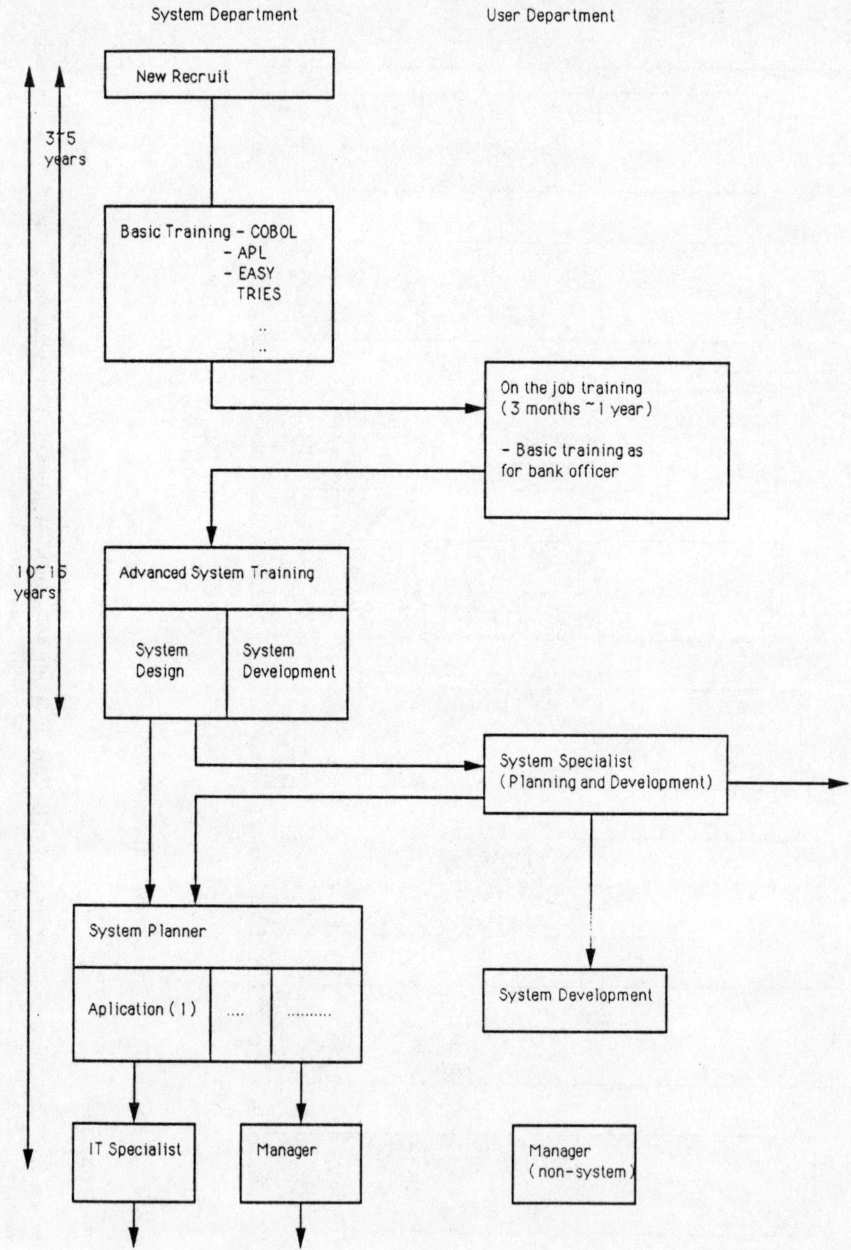


**Figure 5** Career development path of IT specialist

back-office with higher IT literacy may have provided the banks with some strategic options for coping with soaring IT-induced workloads.

*Upgrading human resource at the office*

For the essential purpose of equipping key IT staff for the development process, banks intensify their human resource management. Generally, Japanese firms are known to rely on 'on-the-job' training to improve the quality of their human resources. They design a career path where job promotion (or rotation) results in the accumulation of varied skill and knowledge through on-the-job training in a succession of different posts. And, broadly speaking, the banks are no exception to this general practice (see Figure 5). Except for a few high-grade IT specialists, banks bring on their key people along a rather broad-gauge career track. For instance, a recruit-training package for key-person candidates comprises not only system contents but also the banks' day-to-day operating procedures. Once key people have qualified as system designers, banks tend to rotate them round user departments. Even those at managerial level (e.g. specialising in the design of a new financial product) are not reluctant to move to user departments. Such an interactive career path is designed to create key people of the right sort; those who can successfully link user needs in the office with the technological specifications of the systems.

In the back-office sustaining high morale is no easy matter, since the work environment tends to be somewhat monotonous. This becomes a serious problem with the female workers in particular, who can rarely expect job rotations. Faced with this office predicament, banks introduce multifaceted training (or self-enlightenment) programmes into the back office (e.g. training to select female leaders etc.) with the aim of providing the back-office with targets and some degree of challenge. The back-office welcomes such provisions, and finds them worth while. In addition to the formal training system, we should refer to the total quality control (TQC) in back-offices, which generally came into being in the early 1980s. This movement, too, is designed to give some impetus to back-office morale, but it also yields some practical benefits. It raises the efficiency of the search process in the back-office, leading to a drive for many more profitable ways of IT utilisation. It also facilitates the organisational restructuring process and contributes to raising the productivity of already-introduced ITs.

*Emergent collaborations with external organisations*

Finally, let us widen our perspective to take in the banks' transactions with external organisations. First come the computer vendors (CVs) who collaborate with the banks in designing systems and developing software. In this relationship, the Japanese CVs behave uniquely in providing the banks, as part of the deal, with bank-specific systems engineers (SEs). Once an SE is delivered to a particular bank, he/she remains with them for a considerable period. In the Japanese business scene, where few banks used to have in-house SEs, this service-intensive marketing is warmly welcomed.[6] Banks also collaborate with their affiliated software houses (or some small independent ones) in developing applied software and managing the maintenance of the system. Originally, banks externalised some part of the systems department and made it an affiliate, calculating that such an organisational arrangement would enable them to (1) eradicate promotion (or rotation) problems arising between generalists and IT specialists; and (2) reduce total labour costs by adopting a cheaper pay structure for the affiliate. Service personnel delivered from the affiliate are placed in the main back office, often with longer rotation periods. Such arrangements allow the banks to reduce IT-induced labour costs in a flexible way.

On balance, we can say that the Japanese banks have fashioned a stylised intra-bank arrangement for the effective introduction of ITs into the back office. The overall structure emerging from well-devised collaborations with external organisations allows for the introduction of highly bank-specific systems. It also provides some organisational mechanism to reduce IT costs. On the other side of the coin, banks have shown themselves reluctant to rely on external IT consulting firms as some kind of panacea. In contrast to the position in the US, only a few consulting firms have appeared in Japan.

## The efficacy of Japanese services techniques in the banking sector

### *A comparative illustration of the US banks*

Before proceeding to the main concern of this section, we will schematise a comparative illustration of the US case. In our observation, the stylised US managerial characteristics, job specialisation and exercise of hierarchical control, when introducing ITs were that: profit centres rank candidate plans from the viewpoint of their benefit to the business; systems departments assess the plans on the basis of their calculation of system costs. The final evaluation of the costs and benefits of the plans, and the decision on their respective value, are both made at board level. In making its decisions, a board relies both on computer vendors' proposals and on the reports of external consulting firms. These firms, which guarantee neutrality as between computer vendors, make reference to both the business and system elements of a proposed plan and contribute to the decision-making process at the top. This organisational method gives top management the opportunity for dynamic decision-making leading to strategic IT introductions (e.g. the 24–hour ATM system of Citicorp). It should be pointed out, however, that it lacks any special mechanism for fine-tuning new systems to continuing domestic operations. Unlike the position in Japanese banks, inter-department (i.e. between profit centres and specialised system departments) promotion (or rotation) as a means of knowledge and skill consolidation is of limited significance. The requirement for decisions at the top for any changes actually entails some transaction problems and counter-indications, particularly with regard to incremental introductions and modifications.

### *A fragmentary evaluation of the US banks' IT introduction*

Recognising the strategic value of IT as a competitive weapon, the US banks now invest in the region of $10 billion a year on designing and operating computer systems. The big banks have played a major part in this move. The *Economist* reports that (1) the big banks control around two-thirds of the total spending; and (2) by next year Citicorp may on its own account for 20 per cent of the US banks' spending on computers (the *Economist*, 25 March, 1989). As a consequence of this heavy investment, IT's share of the total capital stock of US banks has shown a drastic rise, having increased from 1.7 per cent (1970) to 10.1 per cent (1980), 27.6 per cent (1985) and 40.4 per cent (1988) (Roach, 1989). Equally, IT endowments per white-collar worker have shown the same trend, raising their average annual growth rates from 3.3 per cent (1963–82) to 18.8 per cent (1983–88) (Roach, 1989). Although no full-scale evaluation of the banks' commitment to IT installation has yet appeared, we can adduce fragmentary evidence pointing to some potential problems.

First of all, as an enormous back-office support staff is put in place to facilitate transactions in a growing banking business, some observers frankly doubt the contention that ITs will raise efficiency in the new 'assembly line' of back-office information production: as Roach suggests, one can ask (1) whether or not IT introductions enable the banks to reduce the back-office content of their white-collar workforce; and then (2) whether or not any resultant reduction can rationalise the heavy outlay on IT equipment, and thus costs. Second, some researchers make a more basic point: that the computer systems the banks have bought or built simply cannot satisfy the banks' job requirements. The *Economist*, for instance, cites (among others) the case of Citicorp's purchase of the Trade Analysis and Processing System from Morgan Stanley. Observation of this case prompts the assumption that malfunction of the system stems partly from the mismatch between a system originally designed for one purpose (e.g. Morgan Stanley's system designed for handling equity trading and settlement) and the operational requirements of a specific job (e.g. Citicorp's job requirements for its foreign-exchange trading). But it should be said that the difference between Japanese and US banks' methods of IT utilisation may be a matter of degree. However, we should explore the comparison further, since an understanding of the characteristics of the US case is beneficial for the discussion which follows.

### *The Japanese services techniques contributing to back-office efficiency*

Thanks to the Japanese banks' systematised efforts to upgrade the quality of their back-office operations (i.e. the change in their recruitment principles and the multilayer training system), a body of multipurpose support staff emerged through the course of the 1980s, with higher IT literacy. In collaboration with the banks' IT specialists, the back offices pushed ahead with a learning process which has gradually formalised a base of system formation knowledge (i.e. about the formation of systems, knowing how to spot and adapt to changing environmental conditions). This type of formation process contributes to the following characteristic features of the banks' IT utilisation.

First, the process reduces the likelihood of mismatches arising between daily operations and formalised systems. It is very subtly devised right from the start to transplant existing operational procedures into systems. Also, when banks buy systems from outside, the process catalyses the original systems, modifying them into customised ones suited to specific on-site daily operations.

Second, the move towards a more intelligent back-office is actually accompanied by organisational restructuring. For the purposes of effective IT introduction, banks encouraged their back-offices to review their ongoing package of operational procedures and organisational structure. Making use of the fruits of the TQC movement etc., banks were then willing to reshuffle their back-office structure to improve the fit between technology and organisation. Certainly, as Ayres describes in the context of the 'information overload phenomenon', 'somewhere in every information processing organization (at any given time) there is a critical human bottleneck . . . When this human bottleneck is overloaded, the organization has production problems, most likely in terms of delivery delays and/or quality control' (Ayres, 1989). So, in Ayres's terms, organisational tuning-up is meant to detect human bottlenecks in the back-office and to get rid of the overload phenomenon in their information processing. This is proved to contribute to (1) reducing peak-time workloads; (2) stabilishing workload fluctuations; and (3) setting a workload standardisation for each operation, etc.

All in all, banks take a systematic approach to IT introduction. They not only manage the technological components of the introduction but also renovate their corporate policies regarding human resources, training, organisational restructuring, etc. so as to provide a new working environment with an IT-oriented back-office. This approach actually results in a process of mutual adjustment between systems and organisation (as we pointed out, a specific organisational setting plays a key role in formalising banks' systems, and the systems introduced into the back-office, in turn, necessitate organisational fine-tuning). The process meshes systems with on-site operational procedures and gives rise to a highly affinitive relationship between technology and organisation. From the viewpont of back-office efficiency, the relationship contributes not only to visible efficiency (i.e. eliminating back-office staff) but also to invisible gains (e.g. re-allocating the excess staff to more productive tasks, etc.).

## Conclusion

Based on our observations so far, we have clarified some aspects of the Japanese services techniques in the management of IT introduction: as described, the approach is very systematic. The use of technology is always reviewed from the standpoints of human resource management, training policy, organisational restructuring etc.; a kind of 'on-the-spotism' contributes to improving the match between technology and organisation. In the case of the banks' IT introduction into routine/domestic operations, the insight leading to an IT introduction derives mainly from ongoing business practice in a bottom-up manner.

However, the banks have now started to meet a new challenge of introducing IT into non-routine areas like building strategic information systems and investment models, etc., or using IT to branch out with new financial products.

Faced with a transformation phase, the banks are attempting a renovation of their corporate strategies: the strategic information department, linked directly to the board, may become an integral part of the standard IT department; in turn, end-users armed with user-friendly hardware and software may develop their own customised systems on site. From the viewpoint of the systematic approach, the new strategic package may be put on a traditional managerial trajectory. For instance, to prepare to meet the increasing demands of the financial technologist, the banks seem to be relying on their traditional in-house training systems. With the aim of raising in-house financial technologists, the banks have started, since about 1987, to recruit a large number of graduates with science and engineering backgrounds. The proportion of science and engineering graduate recruits has now grown to between 10 and 20 per cent of the male graduate recruits (see Figure 3). These recruits tend to follow a generalist career path, with some additional bias towards departments with a high-technology content. Also, if the banks stick to the pragmatic approach, although they are fascinated by the potential value of the strategic ITs, the utilisation process might be counter-balanced by economic calculations. As this implies, this setting need not guarantee any branching out into dynamic IT utilisations for strategic purposes.

In contrast, the US banks' methods of IT introduction can be dynamic: the top management who sometimes have science and engineering backgrounds are accustomed to top-down strategic decision making;[7] research endeavours to improve software development methodology (e.g. rapid prototyping[8]) are much more conspicuous in the US. However, the *Economist* lists up to three types of common malfunctions seen among the US banks so far, i.e. 'banks have failed to find out whether there was a market for the technology-based product or whether the costs of introducing the new system would be recoverable and 'the computer system they had bought or built could not do the job it was required to do'. If we can assume that the malfunctions stem partly from the US firms' presbyopic view of technological potential,[9] cannot we learn from the Japanese services techniques which are basically rather myopic?

## Notes

1. Yasunori Baba is from the National Institute of Science and Technology Policy (NISTEP), Science and Technology Agency (STA) of Japan. Shinji Takai is from the Mitsubishi Research Institute (MRI). The authors are indebted to Satoshi Nakada, Akitoshi Seike and Eric Von Hippel for comments.
2. For instance, in the average annual productivity growth in finance, insurance and the real estate sector, Japan has recorded positive figures (1.86 per cent (1973–85)) in contrast with the US's negative ones (–1.51 per cent (1973–85)) (Gordon and Baily, 1989).
3. Investment level: Y 15 billion; through-put: 1 MIPS; programme steps: 0.5 million; on-line operation: individual accounting (i.e. deposit and exchange); network: intra-bank.
4. Investment level: Y 30 billion; through-put: 10 MIPS; programme steps: 2 million; on-line operation: systematisation of accountings (e.g. among deposit, exchange and financing); network: inter-bank and bank-big firm (i.e. firm-banking).

5. Investment level: Y 50–100 billion; through-put: 100–400 MIPS; programme steps: 5–10 million steps; network: bank-users (e.g. home-banking) and inter sectoral/ international.
6. IBM Japan, which used to set its service contents (e.g. the number of SEs per mainframe computer), seems to have noticed an efficacy of the marketing. IBM Japan has recently started joint ventures with several software houses (e.g. CSK).
7. See, for instance, Reed and Moreno (1986).
8. Rapid prototyping is the process of quickly producing a prototype of a software system according to its requirements. The prototype exhibits the functional behaviour of the target system, although may not meet all its real-time requirements. Use of the prototype provides feedback to the software designers as to the suitability of the system, and gives valuable early experience to future users. Rapid prototyping results in the early establishment of more complete and correct requirements and design (Klausner and Konchan, 1982).
9. As the *Economist* puts it, '[the firms] have been dazzled by the prospects which technologists have dangled before them'.

## References

Aoki, M. (1986), 'Horizontal vs. vertical information structure of the firm', *American Economic Review*, December, pp. 971–83.

Ayres, R.U. (1989), 'Information, Computers, CIM and Productivity', paper presented to the International Seminar on Science, Technology and Economic Growth, OECD, Paris, 5–8 June.

The *Economist* (1989), 'A survey of international banking', 25 March.

Freeman, C. (1987), *Technology Policy and Economic Performance: Lessons from Japan*, London: Pinter.

Freeman, C. (1988), 'Japan: a new national system of innovation', in Dosi, G. *et al.* (eds), *Technical Change and Economic Theory*, London: Pinter.

Gordon, R.J. and M.N. Baily (1989), 'Measurement Issues and the Productivity Slowdown in Five Major Industrial Countries', paper presented to the International Seminar on Science, Technology and Economic Growth, OECD, Paris, 5–8 June.

Imai, K., Nonaka, I. and H. Takeuchi (1985), 'Managing the new product development process: how Japanese companies learn and unlearn', in Clark, K.B. *et al.* (eds), *The Uneasy Alliance — Managing the Productivity-Technology Dilemma*, Boston: Harvard Business School Press.

Klausner, A. and T.E. Konchan (1982), 'Rapid prototyping and requirements specification using PDS', *ACM Sigsoft Software Engineering Notes*, 7 (5) December, pp. 6–105.

Reed, J.S. and G.R. Moreno (1986), 'The role of large banks in financing innovation', in Landau, R. and N. Rosenberg (eds), *The Positive Sum Strategy*, Washington, DC: National Academy Press.

Roach, S. (1989), 'Pitfalls on the "new" assembly line: can service learn from manufacturing?', paper presented to the International Seminar on Science, Technology and Economic Growth, OECD, Paris, 5–8 June.

# 8. International competition and national institutions: the case of the automobile industry

*Ben Dankbaar*

The competitive performance of the Japanese automobile industry over the past two decades has been astonishing. Hundreds of books and articles have been written trying to describe and explain this phenomenon. The car manufacturers of Western Europe and the United States are still recovering from the many blows dealt to them and are anxious to learn whatever can be learnt from Japan. The following chapter deals with the problem of learning from the experiences of others, from other countries with other traditions and institutions. The basic argument is that to learn is not to imitate. Car manufacturers that simply try to copy Japanese practices are unlikely to regain competitiveness in the world market.[1]

## 1. The Japanese 'threat'

In 1988 the world production of motor vehicles (passenger cars, trucks and buses) amounted to 48.6 million units. Of these, 14.9 million were produced in Western Europe. Although there is much talk about a maturing of the car markets, the fact is that the total number of cars on the roads is still growing everywhere. We are all familiar with the consequences: congested cities, air pollution and traffic accidents with a terrible death toll year after year. This chapter, however, deals with another, more mundane, problem that comes to the minds of European car manufacturers upon hearing figures like these. That problem is: why is it that the West European car industry is producing *only* 14.9 million units per year? The European share of world motor vehicle production used to be much higher. Of course, we know what happened: over the past two decades the Japanese motor vehicle industry gained an enormous share of the world market. The traditional homelands of the auto industry were conquered one by one, with the result that approximately every fourth car sold in North America and every eighth car sold in Western Europe is Japanese. The Japanese share of world production was less than 2 per cent only in 1958 (188,000 vehicles). Ten years later it had risen to 14 per cent, in 1978 it was 22 per cent and in 1988 the Japanese manufacturers produced 12,699,803 motor vehicles, or about 26 per cent of world production. About half of these, 6.1 million, were exported (1.4 million to Europe). The stunning competitive performance of this industry alone (but, of course there were more) was sufficient to change the balance in international economic relations, not just between Japan and the United States, but worldwide. It is at the basis of a revival of protectionist sentiment. It also resulted in some of the largest investment programmes that private industry has ever seen, as the old giants of the industry tried to regain strength.

## 2. Explanations

What are the explanations for the success of the Japanese car manufacturers? In the course of time many different explanations have been provided. An early one, probably most current and most valid in the 1960s, referred to the lower wage level in Japanese industry compared to the earnings of the automobile workers in Western Europe and the United States. Even after wages started to rise for the employees of Toyota, Nissan and the other Japanese manufacturers, the argument kept some of its validity due to the low level of vertical integration in the Japanese car industry. The Japanese manufacturers are really assemblers. Parts and components are supplied by a network of independent or semi-independent firms, where wages are usually much lower. Whereas the high level of integration of firms like Ford and General Motors, who even produced some of their own steel and glass, had long been considered a competitive advantage, Japanese practice showed that it could also be a disadvantage, because high 'automotive' wages have to be paid during all stages of the production process. Still, by now the differences in wage levels between Japan and its competitors have become much less important and cannot explain the continuing competitive successes of the auto industry.

Another popular explanation was that the Japanese had simply been extremely lucky. This may be called the explanation of the 1970s. The Japanese were indeed lucky in that they had exactly the right products to offer when the oil crisis hit the West and demand for small, fuel efficient cars soared. Especially in the United States, where cars had become bigger and bigger and fuel prices had always been low, the traditional car manufacturers were caught off-balance. Even if there had not existed a 'big-car culture' in these enterprises, the lead times and organisational rigidities of these huge firms are such that changing their basic orientations takes many years if not decades. The two oil crises of 1973 and 1979 provoked a decade of crisis and uncertainty in the United States car industry. To call the Japanese success pure luck, however, would be too easy. That would not explain why other producers of small cars, who already had access to the United States market (e.g. Volkswagen), could not expand their market share and indeed lost most of it in the course of the 1970s. It was really a combination of circumstances (there was no room for big cars in Japan), foresight (a strong interest in fuel efficiency and pollution control) and luck, plus the willingness and ability to use this golden opportunity. In the end, ability was more important than luck. By 1980 it was clear that the Japanese success was based not just on low wages, let alone on luck, but also on the ability to produce high-quality products at low costs. Several US studies of the Japanese cost advantage, including one ordered by the United Auto Workers, all came to the same conclusion (Flynn, 1982).

High productivity, then, can be called the explanation of the 1990s. High productivity itself, of course, must also be explained. A variety of reasons has been adduced; some already old, others new. One reason was culture. Parts of Shintoism, the recent experiences of war and reconstruction, Japanese group mentality, their willingness to identify with 'their' firm, these and other aspects of Japanese culture were seen to produce a specific Japanese work ethic, comparable to the Protestant ethic that according to Max Weber had explained the rise of capitalism in Europe.

Another reason mentioned was simpler and more straightforward: exploitation. And indeed, the speed and intensity of work in Japanese assembly plants was certainly higher than in most Western plants. It had been described in vivid detail by Satoshi Kamata (1986) in his book 'Japan in the passing lane', based on his own experiences as an assembly line worker. The assembly line, with high-speed, short-cycled production tasks, is still the corner-stone of Japanese industry. Still, it is difficult to explain the competitive advantages of the Japanese only by the fact that the assembly line workers work harder.

A third reason for Japanese performance was found in the levels of mechanisation they had achieved and especially in the use of robotics. Japanese statistics seemed to show that they were way ahead in the use of robots. On closer inspection, the level of mechanisation was less spectacular. Investigating teams from Western countries came to the conclusion that there was nothing in the way of robotics that they did not and could not do themselves. The interesting thing was that the Japanese used robotics and other means of mechanisation where they did not seem to be profitable from a Western perspective.

This finally leads to a fourth major explanation: the Japanese seemed to have a better system of production management. Without denying the importance of the other factors mentioned, the great achievement of the Japanese car manufacturers is no doubt that they have developed new improved methods for the organisation of production in an industry where everything had been improved already over and over again. It is to this point that we will now turn.

## 3. The Toyota Production System

Ever since the days of Henry Ford, automobile production has been organised on a moving assembly line according to the principles developed by Frederic Taylor and by Ford himself. In accordance with Taylor's ideas of 'scientific management', direct production work is short cycled (usually less than one minute) and carefully measured. There is a strong division of labour between work preparation, work execution and quality control. Ford's specific contribution consisted in the introduction of the continuously moving assembly line and other forms of mechanisation, which determined the pace of work. The influence of this production concept has been at work in all manufacturing industries and in all industrialised countries. 'Fordism', as it has been called, was of course also introduced in the Japanese car industry after the Second World War. Schonberger (1982) has even argued with regard to the modern Japanese manufacturing methods, that 'the Japanese out-Taylor us all' (p. 193). In some respects, for instance where it sticks to the central role of the assembly line, this is certainly true. In other respects, however, the Toyota Production System represents a strong departure from the principles that guided Henry Ford.

The Toyota Production System (TPS), which has in its turn become a model for other Japanese car manufacturers and indeed for the world car industry, is connected with the name of Ono Taiichi. With great endurance, if not fanaticism, Ono Taiichi spent several decades at various posts to improve the production system at Toyota (Cusumano, 1985). The guiding principle of his production system is the elimination of waste. Waste is represented, for instance, by in-process inventories (buffers), repair work and workers standing idle. All these are quite common in a traditional Fordist production system, but Ono Taiichi set out to eliminate them. He made direct production responsible for quality, going as far as allowing production workers to stop the line if they cannot achieve the required quality. He eliminated buffers and reduced inventories of parts and components to a bare minimum. Bottlenecks and quality defects thus became visible immediately. Workers were supposed to support each other as soon as they had finished their own particular task (teamwork) and the time allowed for the tasks was reduced accordingly (no more idle workers standing around). Gradually then, the TPS was organised around the two related principles of 'Just in Time' and 'Total Quality Control' (TQC). If everything is produced just in time for the next stage of production and all buffers and inventories disappear, quality has to be perfect (zero defects). Otherwise, the flow of production gets disrupted. But there is more to Total Quality Control than the necessity (and ability) to do everything right the first time. TQC also stands for the idea of continuous improvement of the production system. Workers were not only asked to do everything quickly and correctly, but also to think of ways of doing it quicker and better. Quality Circles were organised where workers could develop new ideas and work them out together with engineers if necessary. This represents another major departure from the principles of Taylor and Ford. Taylor proceeded from the assumption that there is one best way of doing things and on that basis developed 'a fair day's work' as a norm for the production worker. Ono Taiichi's approach is based on the idea that there is always a better way of doing things. The factory can and must be continuously improved. The TPS therefore represents a system of permanent improvement and permanent intensification of work.

## 4. Toyotism

All this resulted in a production system that differs in crucial ways from the traditional Fordist system. Table 1 shows some of the differences by comparing two typical assembly plants in Japan and the United States. Productivity in the Japanese plant (hours per car) is more than twice as high as in the US plant. The total workforce of the US plant is also more than twice as high as the Japanese workforce. If working hours were the same, therefore, both plants would produce the same number of cars! Looking at the structure of the workforce, we find some more characteristic differences. The number of persons working directly in production in the US plant is double that of the Japanese plant. For all other (indirect) functions, however, this relationship is even higher. For

**Table 1** Assembly plant comparison: Japan–United States, *c.*1980 (employees)

| *Function* | *Japan* | *United States* | *Ratio Japan = 1.0* |
|---|---|---|---|
| Quality control | 156 | 359 | 2.3 |
| (inspection) | (120) | (302) | (2.5) |
| (emissions) | ( 26) | ( 37) | (1.4) |
| (engineering) | ( 10) | ( 20) | (2.0) |
| Production control | 95 | 310 | 3.3 |
| (scheduling) | ( 11) | ( 66) | (6.0) |
| (materials) | ( 56) | (216) | (3.9) |
| Product engineering | 22 | 6 | 0.3 |
| Manufacturing | 132 | 411 | 3.1 |
| engineering (maintenance) | ( 62) | (207) | (3.3) |
| (janitors) | ( 10) | (114) | (11.4) |
| Production | 1 324 | 2 640 | 2.0 |
| (painting) | (250) | (421) | (1.7) |
| (assembly) | (641) | (1 603) | (2.5) |
| Management staff | 33 | 132 | 4.0 |
| Grand total | 1 762 | 3 885 | 2.2 |
| Hours per small car | 14 | 31 | 2.2 |

*Source:* Cusumano (1985), p. 329.

Quality Control the ratio is 2.3; for Manufacturing Engineering 3.1; for Production Control 3.3; and the Management Staff in the US plant is four times as large as in Japan. Thus, even though the Toyota Production System started out from the Fordist system, it has now gained a quite distinctive identity and it is only logical that the concept of 'Toyotism' has been coined to designate it as an alternative to Fordism (Dohse *et al.* 1985).

Krafcik (1988a, 1988b) has characterised Toyotism as a 'fragile' production concept as opposed to the 'robust' character of the Fordist system. It is fragile, because it lacks the inventories, buffers, quality control and repair areas that were introduced in the Fordist system to ensure an uninterrupted flow of production. They were considered regrettable, but necessary costs, in view of the limited motivation and lack of knowledge of the assembly workers and the levels of quality achievable by component suppliers. Ono Taiichi saw them as a 'waste' and set out to eliminate them. Of course, this could be done only by eliminating their causes. The functioning and the cohesiveness of this fragile system depends on new, co-operative assembler-supplier relations, on an intricate individualised method of rating and paying employees, on specific labour relations with a management-dominated union organisation. Toyotism is, in other words, not just a new management technique, but involves — as indeed did Fordism — also a great deal of social and institutional innovation.

## 5. The social context

Naturally, the Toyota Production System is rooted in Japanese society. Whatever new methods and organisational forms Ono Taiichi thought up, they had to be fitted to, and accepted by, Japanese social institutions. The development of Toyotism is closely related to the development of economic, social and cultural institutions in post-war Japan. If we look at the present situation, there are some elements in the social context of the automobile industry that are obviously helpful to the success of Toyotism. The industrial relations system in Japan is characterised by strong unions at the company level and very weak union organisations at the industry level. Industrial unionism as exemplified by the United Auto Workers in the United States or the IG Metall in West Germany was defeated in Japan in the early 1950s. With the support of managements, company-oriented unions were set up that were willing to proceed on the assumption that the (long-term) interests of the company are identical with those of the workers (cf. Halberstam [1987] for a description of the struggle for control of unionism at Nissan). Another important element in the context of the industry is the educational system. The Japanese system does not have a vocational training system outside of private industry. After (a generally high level of) general education, workers learn specific trades and professions inside the firm where they come to work. As this training is firm-specific, mobility between firms is difficult, which in turn allows firms to invest in their workers. As a result, the Japanese labour market is highly segmented. There is a large gap (in status and in pay) between those who manage to become members of the body of permanent employees of one of the big firms and those who find themselves on the other side of the track — with less security and lower pay. The economic and social policies of the Japanese government in support of the educational system, of the organisation of the labour market and, last but not least, of the automobile industry itself are, of course, also important elements of the social context of Toyotism. No study of modern Japan is complete without reference to the Ministry of International Trade and Industry. The influence of MITI has often been over-estimated and its plans for the auto industry were often disregarded by the manufacturers. Nevertheless, government industrial policies, protectionism as well as structural policies, the MITI approach towards technological planning and problem solving are all part of the environment that influenced Toyotism.

At the factory level, this environment shows up in various elements of Toyotism that are not immediately visible on the shop floor, but nevertheless are crucial to its success. The permanent employees, for instance, pay a price for their job security and social status where by they are expected to identify completely with their firm, to forgo vacations and be willing to work overtime whenever necessary, to accept new assignments and new locations of work without protest and to devote their free time to think about further improvements in the organisation of their work. They are also expected to accept the intricate system of supervision, rating and rewards, which has led some observers to the statement that the personnel department at the Japanese car plant fulfils the same functions that the industrial engineering department fulfils at the traditional Fordist plant, i.e. ensuring that workers work to the limit of their capabilities. Interestingly, although there may be fewer management staff at the Japanese plant (as shown in Table 1), it was shown in comparison with German plants that there are far more persons with a supervisory function who play an important role in the system of individual rating of performance (Jürgens and Strömel, 1987). The social distance between the workers and these supervisors, who also have direct production tasks, seems to be less than in Western countries. This contributes to the potential for team work. It may be explained by the fact that all workers have entered the firm at the same level and were promoted (or not) through individual ratings and training. Similarly, the social distance between production workers and engineers has remained small due to the fact that the engineers, for lack of staff and colleagues, were forced to make active use of the process knowledge of the workers.

## 6. Learning from Japan

Obviously, the Japanese social context is not available elsewhere. What does this imply for the efforts of Western car manufacturers to learn from Japanese experience and to introduce the Toyota Production System in their factories? Should they try to install not only a new production system but also a new social context in their countries? Is that at all possible? Should they, for instance, try to break industrial unionism or demand the abolition of the independent vocational training systems in their countries? At this point we can roughly distinguish two positions. On the one hand, it is argued that the Toyota Production System is essentially a set of management techniques that can be introduced anywhere in the world. On the other hand, it is said that the success of the Toyota Production System is rooted in Japanese social institutions. A transfer of the system to another social context will therefore be unlikely to succeed.

John Krafcik (1988a, 1988b) has argued forcefully that the social context does not matter very much. He has made a comparative analysis of the performance of auto assembly plants in Europe, Japan and the United States, and sees his evidence as refuting the 'country explanation for high performance'.

> Intra-regional variation in operating performance was found to be significant in North America, Europe, and Japan. Substantial overlap among these regions and relatively consistent international intra-corporation performance supports the notion that corporate parentage is at least as important as location in determining the performance of an assembly plant. [Krafcik, 1988a, p. 115]

Figure 1 shows that of the thirty-eight assembly plants investigated by Krafcik, the best performing plant (in terms of productivity) was Japanese-owned and

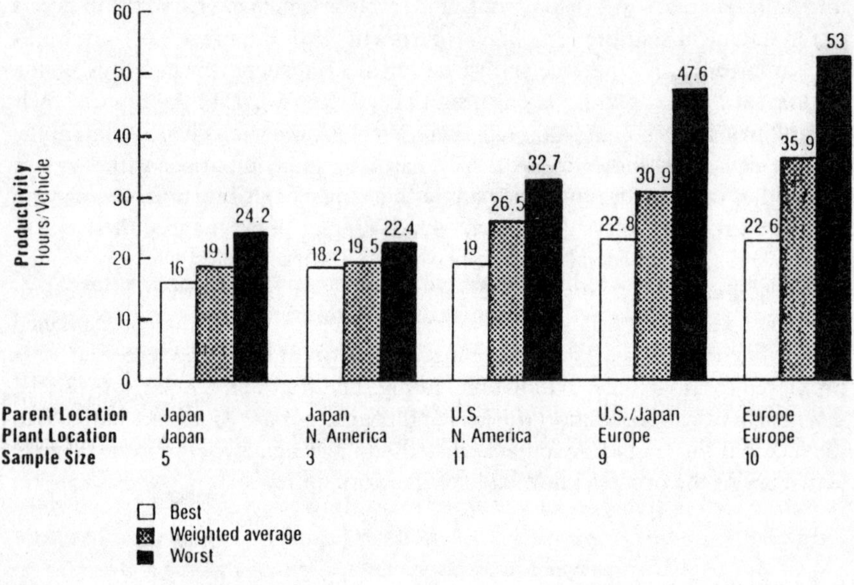


**Figure 1** Range of world productivity performance

*Source:* Krafcik (1988b), p. 46

located in Japan.[2] It also shows that European-owned assembly plants in Europe were on average showing the worst performance. However, the best performing plant in Europe had higher productivity than the worst performing in Japan. Apparently, the 'world class' performance levels of the Japanese plants can also be achieved in other countries. In an effort to explain the variations in performance, Krafcik has developed a composite indicator for the use of Japanese management methods. Included are such indicators as the percentage of floor space available for repair, the presence of teams, the level of unscheduled absenteeism and the possibility for visual control of the production process. He finds that good performance correlates with the use of 'Japanese' concepts of production organisation. Figure 2 shows that 'fragile' management, as measured by Krafcik's indicators, is correlated with high productivity, whereas 'robust' (Fordist) management is generally associated with low productivity. All Japanese-owned plants are in the fragile high-productivity corner. Almost all the plants of one unidentified US multinational, however, are also in this corner. Krafcik cannot say this, but these plants can only be Ford plants. Ford has tried very hard to introduce Japanese methods into its plants. Besides that, it was running almost all of its plants at full capacity during the period under investigation, which in all likelihood led to high productivity scores. Interest-

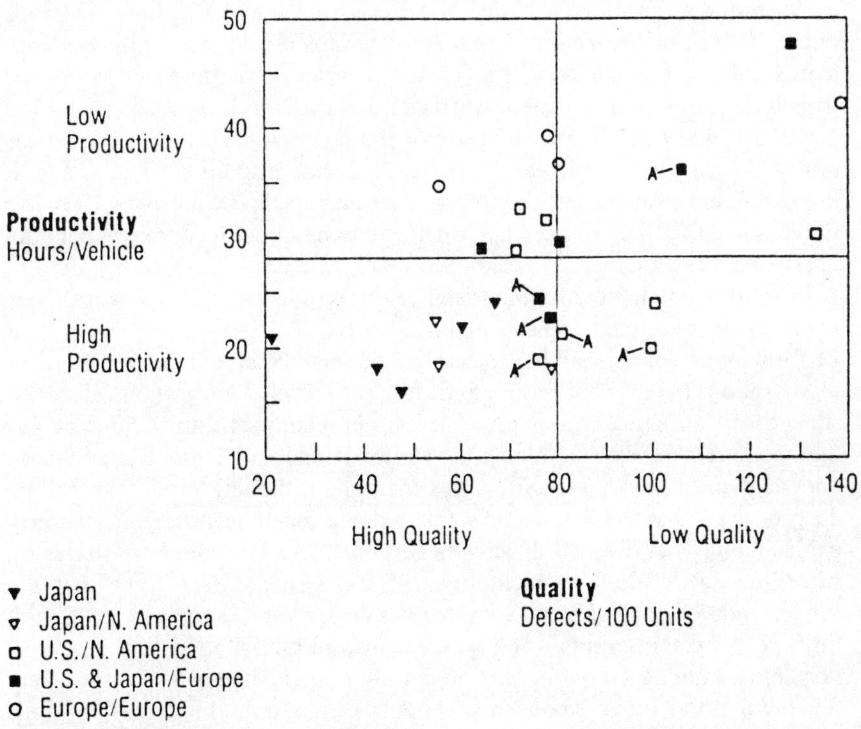


**Figure 2** Management style and productivity: Japan, North America, Europe

*Source:* Krafcik (1988b), p. 49.

ingly, Figure 2 also shows that one of these plants scored high on productivity while the management index indicated a highly 'robust' approach. Krafcik and others who base themselves on the same research (Jones and Womack, 1988), basically argue that 'world class' performance levels can be achieved by adopting the Japanese system of production management. They point to evidence from the United States that it can indeed be transferred to other countries quite easily.

> The opening of the first Japanese plants — Honda in Ohio, Nissan in Tennessee, and the GM/Toyota joint venture in California — quickly eliminated all 'cultural' explanations of Japanese competitiveness. These plants showed that American workers and managers could produce cars with the same productivity and very nearly the same quality as equivalent plants in Japan. [Jones and Womack, 1988]

There is, of course, no reason why the same should not be possible in Europe. And indeed, Nissan has established a new plant in Great Britain, which seems to perform quite well. Ford's efforts, however, to introduce 'Japanese' methods into its British plants seem to be far from successful and meet with resistance from local managements as well as from the workers. Over the past decades, the British car industry has not performed well and the British-owned part of it has performed worst of all. Explanations of the demise of the British-owned car industry have often pointed to inflexible social institutions that made it impossible for the industry to adapt to the new competitive challenges. Krafcik's findings and the Nissan experience, on the other hand, seem to point to a failure of management.

In another international comparison of automobile assembly plants, Jürgens *et al.* (1989) have paid explicit attention to the social ramifications of plant performance. This research project looked mainly at differences in work organisation (team concepts, length of work-cycle, integration of tasks, supervisory structure, automation levels) in plants of three different car manufacturers in West Germany, the United Kingdom and the United States, and tried to explain these differences in terms of country and corporation. Jürgens *et al.* come to a more differentiated conclusion regarding the 'country explanation'. They distinguish between corporate strategies and plant strategies, where the latter are implementations of the former. They argue that the implementation strategies, much more than the corporate strategies, are clearly influenced by the national and sectoral institutions of industrial relations, vocational training, labour market organisation, occupational health and safety regulations and other labour related policies. In fact, differences in work organisation between plants of different firms in one country were often found to be smaller than differences between countries. That does not mean, however, that different corporate strategies are not clearly recognisable in plant practices. Jürgens *et al.* find that the basic thrust of a corporate strategy, for instance whether it emphasises technical innovation or manpower rationalisation, shows up in plant practice of every country, but

> the influence of location of the plant is modifying and overruling many aspects of the influence of corporate decision making. [. . .] The national systems of industrial relations and other labor-related policies and institutions in the three countries each have a particular selective effect on aims and priorities of restructuring activities in the plants. [. . .] These country-specific patterns of selection interact with the corporate strategic profiles, which leads to characteristic blends and mixtures. [Jürgens *et al.*, 1989, pp. 361–3 (author's translation from German)]

Note that these country-specific patterns of plant-level work organisation do not necessarily exclude the possibility of large differences in productivity and product quality between plants in the same country, as they have been found by Krafcik. On the contrary, Jürgens *et al.* find similar differences in plant productivity not only between countries but also within the same country and/ or within the same corporation.[3] The insight that Jürgens *et al.* contribute to the debate, therefore, is not that 'location explains performance', because it obviously does not, but the simple and none the less important point that 'location matters'. The social and institutional context will always be reflected in the actual practice of production organisation. Any effort to introduce new management techniques in a factory therefore will have to deal with this context. Where Krafcik points out that high productivity can be associated with the use of some 'Japanese' methods, Jürgens *et al.* make clear that it is absolutely necessary to find out how these new techniques have been matched to the social context. Whereas a removal of repair areas on the shop floor may be an obvious sign of management's efforts to apply Japanese lessons, management's real achievement in this plant may very well be the way in which it has convinced workers and unions to go along with the new approach. It would then be highly misleading to explain high productivity in this plant by reference to the first measure, which is technically easy to achieve. If management fails to reconcile the new methods and the social context, we can expect either a return to 'robust' management methods or a combination of fragile management with low productivity.

Returning to the questions raised above, we can conclude on the basis of the contributions by Krafcik and by Jürgens *et al.* that the Toyota Production System, brought down to a set of rules for the management of production, can be introduced in other countries. However, just as the TPS is rooted in Japanese society and history, the application of these rules in another country will have to deal with the existing social context, the social institutions and traditions of that country. Just as Fordist factory regimes in France are quite distinct from Fordism in Germany, 'Toyotism' in different countries will also show different features. Able managers can organise high-productivity plants everywhere, however their ability does not show in a slavish application of Japanese precepts, but rather in their handling of the many interfaces between production organisation and the social institutional context.

## 7. Changing social institutions

'Handling the social context' may be easier in some countries than in others. Some important existing arrangements may be very difficult to reconcile with the rules of Toyotism. This may lead managements to the conclusion that the existing institutions have to change or disappear. For instance, some have argued recently that industrial unionism should be eliminated, because it is a relic of Fordism that hinders the introduction of the new world standards. It is important to stand back here, and realise that world market competition does not require firms to copy Japanese production systems. It does require them to realise by and large comparable levels of productivity and quality. Sometimes they may be able to do so simply by adapting the Japanese model to local conditions. Sometimes, institutional changes may be inevitable. In every country, therefore, managements have a more or less wide range of options in dealing with the new competitive challenges, stretching from a 'pure' Japanese production model with completely changed social institutions to a highly adapted production model with almost unchanged social institutions. The actual range of options available to managements in a specific country depends not just on the prevailing context but also on the creative and political capacities of managements.

It is necessary to mention 'political' capacities here, because the process of changing social institutions is necessarily a political one. It involves bargaining with various social groups and organisations, the writing of new contracts, the construction of new institutions and sometimes legislation. Social institutions are usually a product of the interaction of various social actors and cannot be simply created by one group of actors. The interaction can be co-operative, but also confrontational, which again depends not just on circumstances but also on strategic choices made by the actors themselves.

Take, for example, the case of changes in the institutions of industrial relations in the various car producing countries over the past decade. The institutions, rules and organisational forms dealing with industrial relations are typically national in character and vary more between countries than between firms (Dankbaar, 1989a; Jürgens *et al.*, 1989). Over the past decade, managements in all car firms have introduced new technologies and new production concepts. Everywhere this involved dealing with the prevailing system of industrial relations. In Great Britain and the USA, the prevailing system tied workers' rights and rewards closely to the existing 'Fordist' system of work organisation. In the USA for instance, job security depended on seniority, which was tied to a carefully defined system of job classifications. In Great Britain, job descriptions were closely tied to manning levels, with the additional complication that workers with different jobs were often represented by different unions.

In these countries, the introduction of Toyotism, e.g. integrating production tasks and quality control tasks and introducing team concepts, necessarily involved major changes in the balance of interests that had been produced in years of collective bargaining. In Great Britain, managements supported by the Conservative government chose to follow a confrontational course. As a consequence, industrial relations in the British car industry have changed considerably, especially at the plant level, where the power of shop stewards has been reduced. Under these new relations of power, managements hope to be able to introduce new production concepts (Marsden, *et al.*, 1985). The new production concepts, however, require a measure of mutual trust and co-operation that does not fit traditional British industrial relations and the confrontational strategies chosen by the government and managements have only confirmed these traditions. In the USA, car manufacturers negotiated with the United Auto Workers (UAW) and both parties showed an unexpected ability and willingness to try out new possibilities and new contractual arrangements. This has resulted in some major changes in the industrial relations system (wage concessions, a major reduction in the number of job classifications), which could be reached in a much more consensual way than in Great Britain (Katz, 1985).

In both countries we find a tendency to adopt the Japanese model and change the industrial relations system. In countries like West Germany and Sweden, on the other hand, we find a tendency to work with the existing institutional framework and adapt the production model. In West Germany, job security and wages were not closely tied to a Fordist model of production organisation with narrowly defined jobs. Consequently, workers and unions had less reason to consider management's initiatives concerning work organisation as a direct attack on their positions. The changes in work organisation that management considered necessary could either be introduced unilaterally or formally agreed to without much upheaval through the existing bargaining mechanisms. In this respect, the West German industrial relations system turned out to be more flexible than the British or US system. Streeck (1986, 1988) has pointed out that, somewhat paradoxically, this flexibility arises from a system with a strongly formalised structure in which state legislation (co-determination) plays an important role. Legal (status) rights of employees, enforced by the state, make it easier for workers and unions in these countries to negotiate about changes in the organisation of work, even in periods of high unemployment when unions are weak. The absence of such status rights for workers, especially in the field of co-determination, seems to have caused major competitive disadvantages for the car manufacturers in Great Britain and the USA.

Of course, the industrial relations system is only one element of the social and institutional context in which the car manufacturers have to operate their plants. Other important elements are the national systems of education and vocational training, government industrial and technology policies, labour market institutions, traditional attitudes towards work, etc. In view of all these peculiarities of the various production locations, the suggestion that production concepts can be transferred from one country to the other unmodified seems to be highly naive. Managers who nevertheless try to do so, may achieve some improvement in plant performance, but in all likelihood this performance will fall short of the foreign model. They may even be confronted with lower performance levels. The notion that a social context can be adapted at will to a given production concept is, of course, even further removed from reality. In actual practice, both the production model and the social context will change in interaction with each other. It can be argued that it belongs indeed to the essence of entrepreneurship to find a balance between socially acceptable, and therefore effective, changes in the institutional context and improvements in the organisation of work on the shop floor.

## 8. The German/Scandinavian model

In Germany and the Scandinavian countries, efforts to improve productivity and product quality have naturally taken note of the Japanese example, but the willingness of all parties to move within the framework of existing institutions has created the setting for a quite distinctive approach, which is more in line with national institutions and traditions. At the institutional level this 'German/Scandinavian model' is characterised by the existence of legal rights of co-determination for works councils and organised workers, and more generally by a relatively strong involvement of unions and workers at all levels of decision making (Bamber and Lansbury, 1987). It should be added that unions and works councils usually exercise their rights with an equally traditional sense of responsibility and self-restraint. Another important institutional element is the well-functioning system of education and vocational training in these countries, which guarantees a steady supply of well-trained workers.

At the plant level the 'German/Scandinavian model' is characterised by the influence of socio-technical thinking, which is completely absent in the Japanese model. Thus, we find in the German and Scandinavian plants intensive efforts to uncouple labour from the motions of machinery and to introduce stationary work places (usually based on the use of automatic guided vehicles) with longer work cycles, whereas in Japanese plants, the dominant model is short-cycled, machine-paced labour on the continuously moving assembly line. The most extreme example of this movement away from the Fordist assembly line is the new assembly plant of Volvo in Uddevalla, where a small group of workers are assembling a complete car on one spot.

In the German and Scandinavian plants there is a clear tendency to diminish the use of unskilled and semi-skilled production workers and instead have direct production workers, who have completed vocational training in the regular educational system. This can be compared with the situation in Japan, where workers receive quite extensive in-house training. An important difference is the larger possibility for German and Scandinavian skilled workers to move on the external labour market, whereas in Japan workers are largely restricted to the internal labour market. The tendency of rising skills for production workers is widespread in the automobile industry. Motor vehicle manufacturers in the Netherlands, for instance, are following the same line. In co-operation with regular training institutions, they have recently created a new apprenticeship course for automobile assembly workers, which used to be at most a semi-skilled job. Where the regular system of vocational training is less well-developed, firms are forced to develop more in-house training.

The higher level of training is necessary because workers are no longer expected to carry out just one simple task. More and more they are expected to assist each other, to be able to switch frequently to different tasks, to inspect the quality of their own work and even to carry out routine maintenance work. The increasing levels of mechanisation in all parts of the car plant and the consequent rising costs of machine breakdowns and interruptions of production have increased the importance of well-trained and well-motivated production personnel. Frequently, production workers are now organised in teams. In German and Scandinavian plants, however, the meaning of team work differs considerably from the team concept that is prominent in Toyotism. In the European plants teams are meant to be semi-autonomous groups that are able to carry out a specific set of tasks on the basis of their own planning and internal decision-making. In the Japanese plants the teams lack autonomy and function primarily as vehicles for social discipline and control.

## 9. Conclusion

The competitive successes of the Japanese automobile manufacturers have been investigated by all their competitors. Each of them wants to learn 'Japanese lessons'. Especially the methods of production organisation developed at Toyota have attracted attention and the Toyota Production System has become a model for manufacturers all over the world. The argument of this chapter is that efforts to introduce this model in other countries will at best meet with only limited success as long as managements fail to take adequate account of the social context in which their plants are operating.[4] Furthermore, there is no reason to assume that 'world class' standards can be achieved only by following the precepts of Toyotism. In fact, it is one of the basic understandings of Toyotism as developed by Ono Taiichi that there is always room for further improvement. It is therefore within the logic of the model itself that the introduction of Toyotism in other countries will lead to different adaptations and new varieties. The quality of managements in car plants becomes visible, not in their willingness to adopt Japanese methods, but in their ability to apply and adapt these methods in a different social context and to manage the concurrent processes of social and institutional change.

So far, it is only in Germany and the Scandinavian countries that the competitive process in the automobile industry has led to the appearance of a production model with features that differ significantly from Toyotism as well as from Fordism. Like Toyotism this model is rooted in the social context of these countries, and it remains to be seen if it can serve as a model for other countries. In that respect much will depend on the competitive performance of the West German car manufacturers in the coming years, and especially on the performance of the German-owned manufacturers. If they succeed in staving off the present onslaught of the Japanese manufacturers, the German/Scandinavian model may become a source of inspiration for other manufacturers, especially in the other European countries. It is not impossible, however, that these 'German lessons' will be learnt quicker by the Japanese manufacturers than by the Italian, French or British manufacturers.

## Notes

1. Research on which this article is based was carried out in the framework of the following projects: 'Challenges and Opportunities for Employees in the Current Restructuring of the International Automobile Industry (Science Centre Berlin); 'Comparing Capitalist Economies, Variations in the Governance of Sectors'; 'Renewal and Restructuring at Volvo Car BV' (Programme on Technology, Work and Organisation). See also Dankbaar (1988a, 1988b, 1989a, 1989b); Dankbaar and van Diepen (1990).
2. In his analysis Krafcik has made corrections for differences in product, so that his findings express 'real' differences in productivity, as if all plants were producing more or less the same car.
3. Jürgens *et al.* have not corrected for product difference as Krafcik has done, but they have selected plants that are producing comparable models (sometimes the same) for the same market segment. Their data show that, especially in the United States, plant performance is showing strong cyclical movements through the years. Plant comparisons should therefore preferably be based on data over a time period of several years.
4. The fact that Japanese 'transplants' in the United States and Great Britain seem to function quite well, does not necessarily argue against this point. Apart from other

reasons that may explain their productivity (greenfield plants with hand-picked personnel), these plants are most likely to be led by managers who are very much aware of the fact that they are operating in a non-Japanese environment that requires special efforts of adaptation and integration.

## References


Bamber, G. and R. Lansbury (1987), 'Codetermination and technological change in the German automobile industry', *New Technology, Work and Employment*, Vol 2, pp. 160–71.

Cusumano, M.A. (1985), *The Japanese Automobile Industry: Technology and Management at Nissan and Toyota*, Cambridge, Mass: Harvard University Press.

Dankbaar, B. (1988a), 'New production concepts, management strategies and the quality of work', *Work, Employment & Society*, 2 (1), pp. 25–50.

Dankbaar, B. (1988b), 'Teamwork in the West-German car industry and the quality of work', in Buitelaar, W. (ed.), *Technology and Work*, Aldershot: Averbury/Gower.

Dankbaar, B. (1989a), 'Technical change and industrial relations: theoretical reflections on changes in the automobile industry', *Economic and Industrial Democracy*, 10, pp. 99–121.

Dankbaar, B. (1989b), 'Sectoral Governance in the Automobile Industries of West Germany, Great Britain and France', *MERIT-Research Memorandum* 89-008, Maastricht.

Dankbaar, B. and B. van Diepen (1990), 'Vernieuwing en Herstrukturering bij Volvo Car B.V.', *MERIT-Research Memorandum* 90-003, Maastricht.

Dohse, K., Jürgens, U. and T. Malsch (1985), 'From "Fordism" to "Toyotism"?: the social organization of the labor process in the Japanese automobile industry', *Politics and Society*, 14:2, pp. 115–146.

Flynn, M.S. (1982), *Differentials in Vehicles' Landed Costs: Japanese Vehicles in the US Marketplace*, The University of Michigan, Office for the Study of Automotive Transportation, Working Paper Series No. 3.

Halberstram, D. (1987), *The Reckoning*, New York: Avon.

Jones, D. and J. Womack (1988), 'The real challenge facing the European motor industry', *Financial Times*, 28 October.

Jürgens, U., Malsch T. and K. Dohse (1989), *Moderne Zeiten in der Automobilfabrik, Strategien der Produktionsmodernisierung im Länder- und Konzernvergleich*, Berlin: Springer-Verlag.

Jürgens, H. and H.P. Strömel (1987), 'The communication system between management and shop-floor: a comparison of a Japanese and a German plant', in Trevor, M. (ed.), *The Internationalisation of Japanese Business*, Campus Westview, Frankfurt, pp. 92–201.

Katz, H.C. (1985), *Shifting Gears: Changing Labor Relations in the U.S. Automobile Industry*, Cambridge, Mass: MIT Press.

Krafcik, J.F. (1988a), *Comparative Analysis of Performance Indicators at World Auto Assembly Plants*, Massachusetts Institute of Technology.

Krafcik, J.F. (1988b), 'Triumph of the lean production system', *Sloan Management Review*, fall, pp. 41–52.

Marsden, D., Morris, T., Willman, P. and S. Wood (1985), *The Car Industry: Labor Relations and Industrial Adjustment*, London/New York: Tavistock.

Satoshi, K. (1986), *Japan aan de Lopende Band*, Amsterdam: Jan Mets.

Schonberger, R.J. (1982), *Japanese Manufacturing Techniques: Nine Hidden Lessons in Simplicity*, New York: The Free Press.

Streeck, W. (1986), 'The Uncertainties of Management in the Management of Uncertainty: Employers, Labor Relations and Industrial Adjustment in the 1980s', Wissenschaftszentrum Berlin, IIM/LMP 86-26, Berlin.

Streeck, W. (1988), 'Successful Adjustment to Turbulent Markets: the Automobile Industry', Wissenschaftszentrum Berlin, January.

# Part III: Basic theory, theoretical models and policy

# 9. Adoption and diffusion of technology as a collective evolutionary process*

*Gerald Silverberg*

## Introduction

Innovation diffusion occupies a special place in the economics of technological change. On the one hand, it is the best established and most intensively studied empirical phenomenon in this area, and the logistic and other S-shaped curves have provided a sound mathematical, if somewhat phenomenological, inroad into a diverse range of applications. On the other, it remains somewhat divorced from any microeconomically founded overarching theory of the determinats of technological change which might constitute a central component of a general approach to economic dynamics and social evolution.

Why is this the case after many years of diffusion research and a vast accumulation of case-studies and a diversity of mostly *ad hoc* theoretical models? The reason appears to me to be twofold. First, most diffusion research has proceeded on the basis of a number of implicit assumptions which have inevitably narrowed the focus of inquiry and prevented a link-up with other aspects of the economics of technological change. Second, a number of features of the economic history of modern technology which must be obvious to even the most casual observer have failed to gain entrance into the diffusion literature. As I shall argue in the following, these represent two sides of the same coin. Once they are brought together, a new perspective opens up which, in my opinion, may lead to a fruitful generalization and integration of present work.

The study of technical change has been dominated in this century by what I call the linear model, which undoubtedly goes back to Schumpeter (1912): there is a linear progression from invention to innovation to imitation/diffusion. Each of these three stages is distinct, and a technology passes unidirectionally from one to the next, in the course of which the dominant economic factors and the nature of the actors change character significantly. The technology itself, however, remains more or less the same (once it has been invented) — it is simply passed along this pipeline (perhaps undergoing in the process some slight modifications and adaptations) to reveal its economic potential and be exploited until its innovative strength has been exhausted like a squeezed-out lemon. The diffusion literature has naturally embraced an analogous perspective: an innovation arrives at time zero as a consummated creation, like Venus from Zeus's brow. The problem is then to explain why it takes so damn long for it to be completely adopted by its (a priori clearly defined) population of potential

* Forthcoming in *Technological Forecasting and Social Change*.

adopters. Non-adoption is taken as self-evidently irrational, the (Schumpeterian) heroes being the inventors and innovators, the imitators being ambiguous figures obviously necessary to the process but somehow unseemly and undeserving, while a remnant remains of conservative stick-in-the-muds doomed to economic obscurity. The only questions to be examined are what determines the rate of diffusion, what distinguishes early from late adopters and to what extent forecasting (of the rates and ultimate level of saturation) is possible. Implicit in this viewpoint is the notion that the faster the better, the earlier the time-point of adoption the better, the higher the level of diffusion the better.

Another peculiar lacuna in the diffusion literature is the fact that, although the influence of profitability on adoption and diffusion rates has been extensively studied since Mansfield, the inverse — the effect of the timing of adoption on profitability, relative competitiveness and market structure — has surprisingly been largely neglected. In all of the concern to find an economic explanation of technological change, the economic implications of the latter have been left out of the picture, or relegated to the by now familiar implicit assumption that the earlier the better. It is almost as if decisions with respect to technology, although influenced by economic considerations, themselves had no economic repercussions, in particular in terms of the ability to compete. Thus, studies which simply note the number or size of firms which adopt over time overlook the fact that the sizes, profitabilities and strengths of these firms do in fact change, sometimes quite significantly, during the diffusion period, and in part at least precisely due to the adoption decisions implemented. In part, this neglect is justified by the argument that most diffusion processes concern innovations each of which in itself is of secondary importance to the overall performance of a firm. But if this were so universally true, there would seem to be little point in devoting so much scientific attention to questions of diffusion in the first place.

Of course, this characterization is unfair to quite a number of writers on diffusion. The probit and other equilibrium traditions have argued that diffusion proceeds according to a pace consistent with some predetermined distribution of agents' characteristics (e.g. David, 1966; but see Olmstead, 1975 for a counter-argument on precisely David's case-study of the diffusion of reapers which reinforces the thesis of this chapter). Other writers (in particular Rosenberg, 1972; and Sahal 1981) have emphasized that the technical characteristics of an innovation develop simultaneously and interdependently with its diffusion. Yet none of these questions seems to have been dealt with systematically, and above all fitted into the overall pattern of technical change, of which diffusion is only a part, albeit an essential one.

The mirror image of these sins of omission are observations which seem almost too obvious to have to be stated. The point of recollecting them here, however, is that they may serve as an entry into a new perspective on the role of diffusion *per se* in the process of technical change, and thus bring into focus a number of previously disparate elements of the picture. The first observation is that technologies improve and develop considerably during the diffusion process. In fact, in many cases a new technology may actually be inferior in performance to an established one, and only overtake it later (Enos, 1962, for example, gives a number of examples taken from petroleum refining). This is at variance with the implicit assumption noted above that a technology arrives full blown before diffusing, and that development and diffusion can be separated from each other. This adds an element to the adoption decision which until now only the probit models have incorporated, albeit in a very static sense: when to invest in a new technology given that its technical performance characteristics are both uncertain and changing, and that its future potential is poorly known? The fact that, in general, decision-makers are conscious of the changing nature of technologies in their investment decisions and modify their criteria accordingly enables us to connect up with the considerable literature on durable investment under technical change and the role of technological expectations (e.g. Terborgh, 1949 on optimal replacement; and Rosenberg, 1976 specifically on expectations).

But we can go a step further with this observation to make the claim that one reason technologies develop considerably during the diffusion process is precisely due to this process itself. One obvious reason is learning by doing and learning by using on the part of producers and users respectively (there is a large literature on this subject, but the classical references in economics are Arrow, 1962a; and Rosenberg, 1982, Chapter 6). The more a technology is adopted and employed, the faster producers can go down their learning curves due to increased cumulative investment and production, and the better users can exploit it due to their own accumulated experience. Moreover, user/producer interactions play a critical role in the development and adaptation process, as has been pointed out by Lundvall (1988) and von Hippel (1988) in particular. This process of incremental innovation, which quantitatively can be much more significant than the original 'act' of invention, can itself take on the character of invention, not in the sense of the solitary inventor or pioneering firm still secretly cherished by most writers on technology, but what Robert C. Allen has labelled 'collective invention' (Allen, 1983).

Allen defines collective invention as a fourth form of the generation of new practices and technological knowledge alongside non-profit university and governmental research, firm R&D and individual inventors. In this process non-patentable incremental 'inventions', i.e. explorations of possible technological practice, operating conditions, procedures, designs etc., are more or less freely exchanged within an industry:

> The essential precondition for collective invention is the free exchange of information about new techniques and plant designs among firms in an industry . . . Thus, if a firm constructed a new plant of novel design and that plant proved to have lower costs then other plants, these facts were made available to other firms in the industry and to potential entrants. The next firm constructing a new plant could build on the experience of the first by introducing and extending the design change that had proved profitable . . . In this way fruitful lines of technical advance were identified and pursued [Allen, 1983, p. 2].

Whereas Allen restricts his concept of collective invention to a form of technical change in which firms, essentially non-collusively, build on each other's individual experience without committing resources to what we would now call

R&D (and this, he argues, was more significant in the nineteenth century), there are several aspects of the phenomenon which seem to be relevant to a more general evolutionary perspective on technical change in decentralized economies. First, much 'incremental' technical change is incidental to more 'normal' activities of firms such as investment, scaling up and routine problem-solving. Second, technology is to some extent a form of information and thus partakes of some of the peculiar economic properties of this elusive concept. It can be both private and public at different times, tacit or codifiable, reproducible and transmissible at low cost but exploitable only if the recipient has already attained a certain technological level him/herself, and non-additive. However, complementary pieces of information can be synergetic, allowing much larger advances to be achieved when combined than if they were used in isolation.

In dynamic terms these properties are the preconditions for the well-known Schumpeterian interaction between innovator and imitator. The implicit (linear) assumption of this picture is that the innovator enters with the new information, which eventually leaks out and is copied by others, enabling them gradually to catch up with his superior technological level and whittle away his superprofits. In view of our brief discussion of collective invention above, however, we may be emboldened to break with this one-way concept of innovation causality and advance the hypothesis that imitation and diffusion themselves contribute significantly to the further technological maturation of the original innovative idea through a complicated, two-way process of interaction and collective exploration. Almost any economic history of the development of technology which goes beyond the Samuel Smiles's exaltation of the heroic inventor will confirm that most innovations pass through many hands and strang byways before they attain the 'dominant design' by which we now know them. Rare is the example of the Edison light bulb and electric power system which was single-handedly developed to market maturity by one research laboratory under one man's direction (cf. Friedel and Isreal, 1986). But even here, the original idea is much older, Edison's original inspiration (an electro-mechanical solution based on telegraph technology) was a dead-end, and crucial ancillary equipment (such as vacuum pump technology) came from outside. Furthermore, the important further advances on this technological trajectory, such as high-voltage AC generation and transmission, came about through a multitude of partly complementary and partly competing international contributions totally outside of Edison's control and against which he actually fought a vain rearguard action for a time (see Hughes, 1983; David and Bunn, 1988).

Many of these factors and feedback channels have been recognized in the literature, though usually in isolation. From the economic point of view, the discussion has focused on their influence on the *appropriability* of technical change to the individual agent, that is, the agent's ability to capture a private economic reward for his inventive exertions and exclude free-riders. The information character of technology is what puts this appropriability in doubt, of course. The classical solution is the legal institution of the patent. But it has become clear in recent years that patents are suitable protection for only certain classes of innovations. And in areas of very rapid technical change and concomitant rapid obsolescence of ideas, secrecy or first-mover advantages can be more cost effective forms of protection than the time-consuming and expensive process of patenting, which moreover mandates disclosure (cf. Levin, Cohen and Mowery, 1985). Furthermore, patents can often be designed around; another example of the peculiar forms information 'interchange' can take. The appropriability issue, even in a neoclassical setting, also leads to the well-known justification for government-sponsored R&D due to the discrepancy between social and private rates of return on R&D investment (Arrow, 1962b).

Appropriability is a fascinating issue for the economist because it is an example of an externality, and thus poses a challenge to the optimum welfare implications of those styles of general equilibrium analysis which, at least in theory, fully reconcile individual and social interests. This has led to a sophisticated literature on whether too much or too little R&D will be done compared to some posited social optimum, due either to the inadequate private incentive or to the danger of redundancy and duplication of research efforts. From the perspective of this chapter I think these questions are somewhat beside the point. As I shall try to demonstrate in the following, both externalities and (near) duplication can be very useful, perhaps even necessary, components of technical change when see as a collective evolutionary process. The key concept in this regard is *learning*, which can take place within the individual, the organization and collectively through a network of feedbacks unfolding over time between both co-operative and competitive agents.

## A self-organizational approach to the modelling of technology diffusion

I want to draw on the above observations to indicate some of the features of an appropriate modelling framework for the analysis of technical change, based on some of my own previous work (Silverberg, 1987; Silverberg, Dosi and Orsenigo, 1988; Silverberg, 1990; and the literature reviewed in Silverberg, 1988, 1989). To begin with, it is clear that if technical change is an ongoing process, albeit proceeding in fits and starts, then the decision procedures of participants must eventually take this into account, and thus cannot be rooted in purely static analyses. This is one of the chief reasons (though certainly not the only one) in my opinion for discarding the production function approach for anything but rough indexes of total factor productivity (cf. Silverberg, 1990). In particular, investment and choice of technique decisions must be predicated on expectations about the future rate at which present commitments become technologically obsolete. Fortunately, there exists a considerable literature under the name of optimal replacement theory dealing with precisely these questions, if under certain simplifying assumptions, and insufficiently known to most students of technical change (Terborgh, 1949; Massé, 1962; Smith, 1961; Malcomson, 1975). The essence of a careful optimization treatment of the replacement decision is that the most common rule of thumb in practice, the cut-off (undiscounted) payback period, is theoretically valid, with the payback period determined by the long-run (expected) rate of future technical change and, to second order, the cost of capital.

This conclusion is only really valid, however, for fairly routine forms of capital-embodied incremental technical change and abstraction from problems of market power strategic rivalry. Given that entrepreneurs do employ the payback criterion (though with somewhat varying values of the required payback), one may ask how they collectively learn what the correct payback for any given historical epoch of technical change should be.[1] To this end I have constructed a simple evolutionary model of market competition and incremental technical change which is dynamic, rooted in bounded rationality and plausible decision rules, and displays the properties of cumulative causation and selection (Silverberg, 1987).

One pathway of diffusion is already incorporated in this kind of model: the vintage effect (whose pedigree goes back to Solow, 1959; Salter, 1960; and Kaldor and Mirrlees, 1962). A rational investor will only gradually replace installed equipment once the gain in performance satisfies certain criteria.[2] This does not explain, however, why different agents first adopt a new technology at different times, only why diffusion through the aggregate stock of 'machines' takes time. An obvious reason is that different entrepreneurs employ different payback periods, as many surveys of business practice have repeatedly confirmed. Moreover, their assessment of cost savings and the appropriateness of a technology to their line of business may differ (much as in the probit-type models). The variance in payback periods is not unexpected, given that these should depend in the first instance on expectations about the (necessarily uncertain) rate of increase of the cost advantage of new techniques over old ones, as well as the cost of capital. But it is also clear that systematic differences exist in the customary values in general use in different countries (see Silverberg, 1990), something which probably cannot be explained by reference to differences in the national costs of capital (about which there is considerable definitional and measurement confusion in any case) alone, as Flamm (1988), for example, argues.

If we assume that the rate of change of best practice productivity is exogenously given, then the 'optimal' payback period will allow a firm to track technical change by continually incorporating new equipment into its capital stock at a rate which just balances average unit cost advantages against the financial costs of capital turnover due to acquisition and scrapping. In contrast to the usual derivations of optimal replacement policy, however, this model critically turns on the dynamics of oligopolistic competition. Experience with simulation experiments seems to indicate that a unique 'optimal' payback period may not exist independent of market structure and pricing strategies (as mark-ups on unit costs). In particular, collective effects may be at work which may lock an industry into alternative combinations of locally 'evolutionarily' stable payback periods and mark-ups. This is the focus of my current research and is a topic I cannot go into further here.

Since the thesis of most recent work in the economics of technical change is that the creation of best practice technology itself must be seen as endogenous to the system, this sort of model can be only a stepping-stone to further work. This is particularly the case given that the thesis of the present chapter is that this process is endogenously intertwined with, and inseparable from, the process of diffusion as well. Thus the question is: how are we to model these feedbacks, and at what level must they be situated?

The concept of technological trajectory (Dosi, 1982; Nelson and Winter, 1982; Sahal, 1985) comes to our rescue at this point, enabling us to generalize the previous model in the direction of bona fide diffusion. As Grübler (1989) points out:

> should we not attempt to define the object of diffusion research prior to analysis by some sort of 'evolutionary tree', spanning the whole diverse domain into which any particular case is embedded? Improvements and add-on innovations, which result that a particular technology would become competitive within a given market segment, could be represented as 'branchings' along an evolutionary 'tree' and allow for classification, taxonomy and rigorous definition of technologies in competition.

A practical example of the utility of this 'morphological' approach is also provided by Foray and Grübler (1989) on casting technologies. Technological trajectories enable us to distinguish, at some hierarchical level, between true branchings and incremental advances along well-defined technological 'chreods' (to borrow a biological metaphor from Waddington, 1976). But what does a true branching, which seems to be what we really want to focus upon in diffusion studies, imply for the economist, as opposed to everyday technical change?

A change of technological trajectory entails, first of all, a quantum jump in uncertainty, not so much with respect to the relative merits of competing technologies at a given time but rather concerning the rate and extent of future developments, since extrapolations from past exiernece, which may have been specific to the old trajectory, lose their validity. (In fact, even the further development of the old trajectory may be radically influenced by the advent of the new one, as the often remarked sailing ship effect seems to indicate.) Furthermore, the nature of these developments may well depend on the expectations and commitments of others, i.e. have a strong bandwagon element, such as has often been observed about the standards setting problem and network externalities (cf. Arthur, 1988).

We can go a step further, however, by recognizing that technologies are not just blueprints or collections of artefacts, but also require complementary skills, tacit and codifiable know-how, and proficiencies of agents at various levels of involvement with them, from management and engineers to foremen, skilled and unskilled workers. These skills, organizational structures and the like may often be highly specific to a particular trajectory (one need only think of the myriad organizational differences between job-shop skilled worker production and the Fordist assembly line). They are created in ways which are still poorly understood, as a generalized form of individual learning so to speak, but with specific team, organizational and cultural components which still very much elude scientific analysis. Phenomenologically we can often represent them by means of the familiar learning-by-doing power law relationships. The other side of this coin is the fact that learning which takes place in one organization can 'leak out' (or be stolen) to the benefit of another. Crucial in this regard is the dynamic aspect: the rates at which skills can be acquired through own activity versus the lags in their more general diffusion through formal and informal networks of communication and exchange.

Whereas the adoption/investment decision could be reduced to an application of the simple payback method in the previous model (albeit with some uncertainty as to the appropriate payback period to use), adoption decisions with respect to genuine changes of technological paradigm present some novel aspects which may generalize to other instances of the basic problem of innovation. First, decisions in the present have to be made on the basis of no more than hunches about the development potential of a technology in the future. This is as true of invention as of the adoption of a rapidly changing new technique (such as numerically controlled (NC) machine tools or computer-aided design (CAD)). We may even view these as being phases along the path from idea to actuality in which dimensions of uncertainty are reduced while the standard design(s) gradually comes into being. Second, the presence of both internal and inter-agent learning leads to the following dilemma. Should one adopt early/commit R&D to pre-emptory innovation, in order to secure a competitive lead (or, in the ideal but rare case, a watertight patent) via technological accumulation internal to the firm? Or should one keep one's powder dry, so to speak, and wait for the technological smoke to clear a bit (primarily due to the resource-intensive efforts of others in diffusing and developing) before entering the fray? Kleine (1983), Rosenberg (1976) and Stoneman (1976) provide good examples of this reasoning from different industries. Cohen and Levinthal (1989) go a step further and argue that much R&D itself is not undertaken primarily to realize own innovations so much as defensively, in order to allow the firm to keep up with the progress of others in the field.[3] This confirms once again my hypothesis that R&D and innovation on the one hand, and adoption/diffusion on the other have very many more structural features in common than has usually been remarked. The hard and fast Schumpeterian distinction between innovation and imitation may thus have outlived its usefulness.

In Silverberg, Dosi and Orsenigo (1988) these stylized facts find expression in a simple but dynamic generalization of my model of incremental technical change. The productivity of a vintage of new technology at any point in time is posited to be the product of an underlying embodied maximum value and the efficiency at which it can currently be exploited within the economic organization due to cumulative learning. To each technological trajectory is associated a skill level internal to the firm and a 'public' skill level available to new entrants, which lags behind the average of internal levels and results from the spillover externality. Firms express their assessment of the relative advantages of early entry into the new trajectory by augmenting their standard payback period by some factor, which I term the anticipation bonus.

On the assumption of a diversity of firm anticipations it can be shown that the *dynamic appropriability* of an innovative strategy is very much a function of the rates of learning, both internal and public. Thus, for high enough values of the rate parameters governing each form of skill accumulation, first movers do derive the largest net benefit, in terms of ultimate gains in market share, from the introduction and diffusion of the new technology. This is, of course, the classical

Schumpeterian picture. For intermediate values, however, a second mover or imitator may be able to capture more of the benefits by letting early innovators bear an excessive share of the development costs. For even lower values of these parameters, an innovation will fail to reach maturity and its diffusion will spontaneously reverse, even though some firms are willing to commit to it and it is indeed potentially superior. Given that an unbiased survey of innovation prospects would seem to show that a large if not major share of all innovations fail to diffuse successfully, it is reassuring to find a theoretical model which can also yield this result, though this eventuality did not play a role in its original formulation.

Why, then, do firms differ in their assessment of, and commitment to, a new technological trajectory, and what implications does this fact have for the rate and direction of technical change? First, our results, while superficially paradoxical, are really self-consistent. Firms do not know whether in any particular case a 'first-in' or a 'wait-and-see' attitude is superior in terms of appropriating permanent gains from technological investments. Those that are sufficiently aggressive will secure 'first-in' status, but at the risk of not having the staying power to see an innovation to maturity (on the assumption that it indeed does have potential). But even in this worst case for them, they may serve as the essential trigger to subsequent adoption decisions of other firms who do eventually see the innovation through to profitability. This is by no means uncommon in the history of technology, as Marx already remarked. The lure of the (Schumpeterian) innovative laurels is what sparks this process in sufficiently entrepreneurial economies, but the model provides no rational guarantee for the

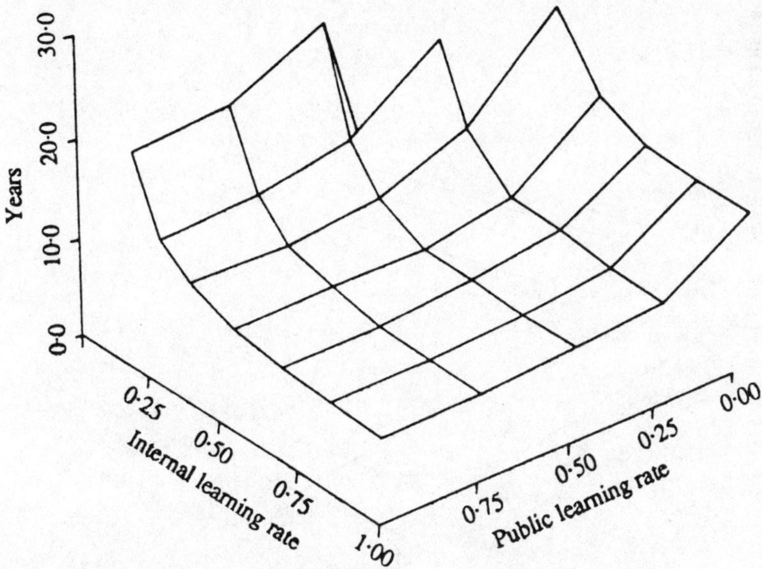


**Figure 1** Time to diffuse from 10 per cent to 90 per cent of industry capacity as a function of the internal and public learning rates

appropriateness of this stance. Finally, too conservative an adoption strategy may lead to the inexorable elimination of the firm from the market and preclude any possibility of ever catching up, due to the cumulative causative forces at work in the model. Thus, a certain amount of innovativeness becomes almost compulsory for everyone once it has become at all widespread among competitors.

The mesolevel manifestations of these contending considerations are also quite intriguing. If we plot diffusion speed (measured, for example, as the time to go from 10 per cent to 90 per cent of the saturation share) as a joint function of these two learning rates, then we find that they *both* contribute to more rapid diffusion (see Figure 1). Moreover, for a given rate of internal learning, an increase in the rate of 'leakage' or public learning does not undermine the appropriability of an innovative strategy, as one might expect, provided one is indeed already in the Schumpeterian regime (see Figure 2). Thus, higher rates of public learning are always socially desirable if the rate of internal learning is sufficiently high to guarantee economic incentives to innovation, if only risky ones.

This state of affairs confronts innovation and industrial policy with a curious, two-handled instrumentarium. On the one hand, at least in a private capitalistic economy, it is necessary to ensure that technological investments are at least in part (and then only temporarily) appropriable. On the other, it is highly desirable to encourage information flows, quick diffusion of experience, and formal and informal co-operation and sharing. The example of collective invention investigated by Allen reveals the spontaneous, one is tempted to say self-

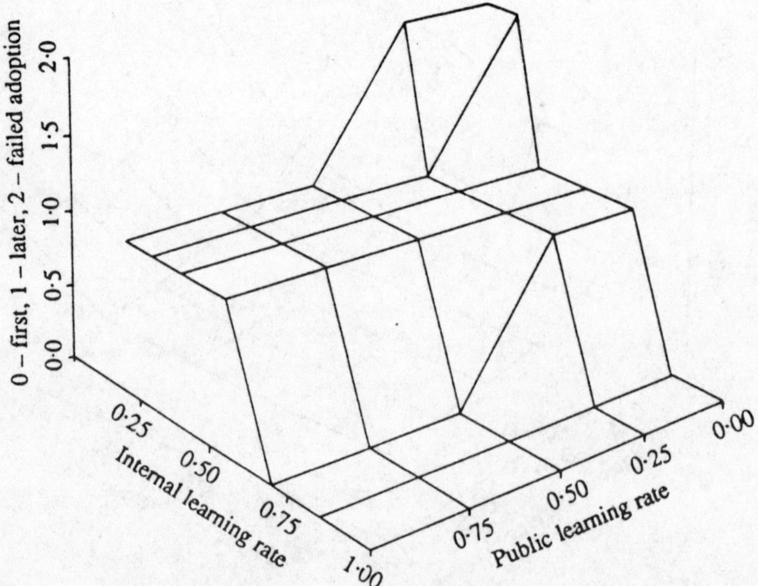


**Figure 2** Diffusion winner: first verses later adopters, for a given distribution of strategies, as a function of the learning rates

organizational, forms this can take which at first glance seems to run contrary to the principles of economic self-interest. An industrialist would gradually extend the technological frontier by building blast furnaces which were slightly taller, slightly larger and operated at higher blast temperatures. This clearly produced fuel economies, but too large jumps along this technological trajectory often resulted in conspicuous failures. By pooling the accumulated experience of different furnaces it was possible to avoid mistakes and channel efforts in the right directions. This was possible because engineers quite willingly published and exchanged operations data on new furnaces. Allen rightly asks why firms would disclose this highly useful proprietary information to competitors and potential entrants. In this case they seemed to have little to lose from disclosure, a bit of prestige to gain from technical publicity and significant mutual benefit to be derived from collective invention in terms of reaping production returns in relation to other steel-producing areas. Von Hippel (1988) also provides contemporary examples of engineers from different, sometimes competing, firms informally and unofficially exchanging information.[4] Saxonhouse (1974) relates how Japanese textile manufacturers pooled operating and equipment experience through their trade association during the early industrialization period. All of these examples hinge on the quid pro quo nature of the ongoing interaction between the agents which is reminiscent of the kind of emergent informal co-operation analysed by Axelrod (1984). From a myopic rationality standpoint, this may appear paradoxical, but seen in the long term, when agents repeatedly interact and come to terms with one another, it may well be in the interests of competing agents to engage in certain forms of tacit co-operation and give and take. Industrial policy has begun to recognize this fact by encouraging and legalizing certain forms of precompetitive joint R&D, such as is exemplified by Sematech and other projects in the USA. The Japanese appear to have been singularly successful in reconciling economic competition with R&D co-operation and technology targeting.[5] Moreover, this synergy is potentiated when the rates of both forms of learning are high, for diffusion of knowledge and co-operation alone is relatively ineffectual. It is necessary for it to be accompanied by a correspondingly high ability to assimilate and internally storm the internal learning curves of the firms themselves in order to reap all the benefits. Thus the appropriability issue does not necessarily lead to a contradiction between public and private exploitation of technological innovations, for they can still be made socially and privately compatible if the learning rates are *both* sufficiently high.

If we want to descend from the Olympian heights of this sort of analysis to the actual determinants of what we have labelled public and internal learning, we are forced to confront a wide range of sociological, institutional and even cultural issues. Thus a major difference between the innovation systems of the USA and Japan relates to the very different (wo)manpower, training and employment traditions of the two countries. The turnover of skilled personnel in the USA is quite high, it being quite common for even top executives and engineers frequently to take jobs with rival organizations. This has been seen as a certain disincentive for firms to invest in human capital formation in the USA, since they are not assured of the 'appropriability' of this investment in terms of employee loyalty. In the context of our model, this could be interpreted as lowering the rate of internal learning. Japan, on the contrary, with its system of lifetime employment, at least for the leading export firms, has evolved a unique system to build up internal skills, loyalty and shop-floor co-operativeness. There are, of course, two sides to this issue, the high mobility of entrepreneurially minded labour in the USA leading to frequent innovative company spin-offs (the semiconductor and computer industries being prime examples, where dissatisfaction with established companies' technology policies has led to several generations of new foundings). While this may increase the rate of public learning, it may not guarantee a sufficiently high rate of internal learning (which, as we have seen, is quite crucial) to be socially desirable in terms of firm and national levels of technological appropriability. In particular, the Japanese may be enjoying the best of both worlds due to their very successful system of pre-competitive co-operation and co-ordination.

Henry Ergas (1987) has introduced the distinction between innovation- and diffusion-oriented technology policies. While I do not quite find his choice of labels apposite (as Giovanni Dosi has suggested, mission-oriented versus generic broad-based might be more appropriate), this distinction is not without relevance for our purposes. A case in point is numerically controlled machine tools, a major innovation dating back to the early 1950s, and the desire of the US Air Force to automate certain involved operations in aircraft manufacture.[6] The technology which resulted was indeed very sophisticated, in fact too sophisticated for anything but subsidized government work. It was not until over a decade later that the technology was increasingly tailored to the needs of general industry and became economically viable, but then US firms, which had specialized in the expensive and sophisticated varieties, were no longer in the forefront. This example reinforces my contention that innovation and diffusion are never entirely separable. The filling of different market niches, itself a form of diffusion, is very often contigent on a sequence of subsequent innovations, both major and incremental. The casting example analysed by Foray and Grübler (1989), in which a new process eventually spread from the batch to the mass production segment of the industry precisely because of the learning economies realized during the initial diffusion phase, is also in this vein. A technology policy that does not take the interdependence of these two aspects into account will always be inherently flawed.

## Theoretical outlook and conclusions

The concepts employed in our model have many points of tangency with other modelling approaches in the literature. The question of internal learning and spillovers in R&D and diffusion has also been examined in a neoclassical, equilibrium framework by Spence (1984) and Jovanovic and Lach (1989), for instance. While I cannot go into the technical issues which separate our approach from theirs here, I would only point out that we come to much less pessimistic conclusions regarding the disincentives and limitations to appropriability than they generally do for systems in the *high* internal learning regime. This is due primarily to the more dynamic, disequilibrium formulation for which we have opted, which represents innovation advantages as being only temporary but potentially leading to cumulative divergences in development paths.

Our work, of course, is very much in the modelling tradition established by Nelson and Winter (1982) and further extended by Winter (1984) and Iwai (1984a, 1984b). An important distinction is the insistence on the embodied nature of much technical progress, i.e. that it can only be realized when new investment is taking place. This has not been emphasized sufficiently in the past in evolutionary modelling (although it should not be over-emphasized, as in the pure vintage modelling tradition). Investment can influence the rate of technical progress through yet another pathway, however. If innovation is collective in the sense of Allen, then further exploration and extension of the technological frontier will only take place if new investment is kept up on a broad front. Allen argues that the demand stagnation of the last quarter of the nineteenth century in Britain brought investment, and thus this form of productivity-enhancing technical change, to a halt, which eventually led to an undermining of Britain's technological competitiveness and a further fall in its world market share.

The endogenization of the innovation frontier is the pre-eminent need in this area. I believe new analytical constructs will have to be introduced to deal with this problem, since, contrary to conventional wisdom, true invention and innovation are more analogous to a semi-blind, semi-stochastic exploration of a rugged landscape than an optimization problem with well-defined choice sets and pay-off matrices. The application of analytical and simulational tools from mathematical biology and artificial intelligence, such as genetic algorithms, evolutionary stable strategies and classifier systems, although bearing the danger of misplaced analogy, may hold the key to further progress.[7]

## Notes

1. This assumes of course that they have in fact discovered the 'correct' value. One could argue, however, that these technological expectations might under certain circumstances be collectively self-fulfilling, so that a multitude of (possibly) suboptimal technological regimes could exist. This is the object of current research I am conducting which I cannot go into here.
2. The simple criterion that total (current plus amortized capital) costs of the new investment equals current costs of the old is widely applied in the literature. Thus, the very provocative discussion of the rationality of late Victorian entrepreneurs is usually couched in this framework (cf. McCloskey 1974a; Allen, 1981). This is all the more remarkable as this is definitely not the 'rational' solution to this problem, as a glance at the optimal replacement literature would immediately reveal. We seem to be confronted here with another instance of academic economists presuming to have a claim to superior rationality over the practitioners (whose rationality, ironically, they are attempting to vindicate).
3. Thus they write (pp. 569–70):

> we argue that while R&D obviously generates innovations, it also develops the firm's ability to identify, assimilate, and exploit knowledge from the environment — what we call a firm's 'learning' or 'absorptive' capacity . . . the exercise of absorptive capacity represents a sort of learning that differs from learning-by-

doing . . . Learning-by-doing typically refers to the automatic process by which the firm becomes more practiced, and hence, more efficient at doing what it is already doing. In contrast, with absorptive capacity a firm may acquire outside knowledge that will permit it to do something quite different.

4. I am very grateful to Harvey Books for bringing this example to my attention.
5. Thus Flamm (1987, p. 151), for example, on computers:

> Joint research, a major element in the rapid development of Japanese computer technology, has created a unique mix of cooperation and competition. In general, Japanese authorities have worked to preserve competition in 'downstream' applications and commercialization of new products. But the results of more basic, precompetitive joint research have been shared quite widely to eliminate wasteful duplication and increase productivity of R&D spending.

6. See Noble (1984) for a somewhat polemical and contradictory but valuable account.
7. On genetic algorithms, see Goldberg (1989) and Holland (1975), on classifier systems, Holland, Holyoak, Nisbett and Thagard (1986). One application pathway to technology has been explored by Kwasnicka and Kwasnicki (1986) and Kwasnicki (1989). The original Nelson and Winter model, of course, can also be viewed as a kind of evolutionary algorithm which tries quite deliberately to argue by economic and not biological analogy.

## References


Allen, R.C. (1981), 'Entrepreneurship and technical progress in the northeast coast pig iron industry: 1850–1913', in Uselding, P. (ed.), *Research in Economic History*, Greenwich, Conn.: JAI Press, pp. 35–72.

Allen, R.C. (1983), 'Collective invention', *Journal of Economic Behavior and Organization*, 4, pp. 1–24.

Arrow, K. (1962a), 'The economic implications of learning by doing', *Review of Economic Studies*, 29, pp. 155–73.

Arrow, K. (1962b), 'Economic welfare and allocation of resources for invention', in Nelson, R.R. (ed.), *The Rate and Direction of Inventive Activity*, Princeton: Princeton University Press.

Arthur, W.B. (1988), 'Self-reinforcing mechanisms in economics', in Anderson, P.W., Arrow, K.J., and D. Pines (eds), *The Economy as an Evolving Complex System*, Reading, Mass.: Addison-Wesley.

Axelrod, R. (1984), *The Evolution of Cooperation*, New York: Basic Books.

Cohen: W.M. and D.A. Levinthal (1989), 'Innovation and learning: the two faces of R&D', *Economic Journal*, 99, pp. 569–96.

David, P. (1966), 'The Mechanization of reaping in the ante-bellum midwest', in Rosovsky, H. (ed.), *Industrialization in Two Systems: Essays in Honor of Alexander Gerschenkron*, New York: Wiley.

David, P. and J. Bunn (1988), 'The Economics of Gateway Technologies and Network Evolution: Lessons from Electricity', paper presented at the International Schumpeter Conference, Sienna, 24–28 May.

Dosi, G. (1982), 'Technological paradigms and technological trajectories', *Research Policy*, 11, pp. 147–62.

Enos, J.L. (1962), *Petroleum Progress and Profits*, Cambridge, Mass.: MIT Press.

Ergas, H. (1987), 'Does technology policy matter?', in Guile, B.R. and H. Brooks (eds), *Technology and Global Industry: Companies and Nations in the World Economy*, Washington, DC: National Academy Press.

Flamm, K. (1987), *Targeting the Computer: Government Support and International Competition*, Washington, DC: The Brookings Institution.

Flamm, K. (1988), 'Differences in Robot Use in the United States and Japan', paper presented at the OECD workshop on 'Information Technology and New Economic Growth Opportunities for the 1990s', Tokyo, 12–23 September.

Foray, D. and A. Grübler (1989), 'Morphological Analysis, Diffusion and Lock-Out of Technologies', paper presented at the International Conference on Diffusion of Technologies and Social Behavior, Laxenburg, 14–16 June, forthcoming in *Research Policy*.

Friedal, R. and P. Isreal, with B.S. Finn (1986), *Edison's Electric Light: Biography of an Invention*, New Brunswick, NJ: Rutgers University Press.

Goldberg, D. (1989), *Genetic Algorithms in Search, Optimization, and Machine Learning*, Reading, Mass.: Addison-Wesley.

Grübler, A. (1989), 'Diffusion: Long-Term Patterns and Discontinuities', paper presented at the International Conference on Diffusion of Technologies and Social Behavior, Laxenburg, 14–16 June, forthcoming in *Technological Forecasting and Social Change*.

Holland, J.H. (1975), *Adaptation in Natural and Artificial Systems.* Ann Arbor: University of Michigan Press.

Holland, J.H., Holyoak, K.J., Nisbett, R.E. and P.R. Thagard (1986), *Induction Processes of Inference, Learning, and Discovery*, Cambridge, Mass.: MIT Press.

Hughes, T.P. (1983), *Networks of Power: Electrification in Western Society 1880–1930*, Baltimore/London: John Hopkins University Press.

Iwai, K. (1984a), 'Schumpeterian dynamics; an evolutionary model of innovation and imitation', *Journal of Economic Behavior and Organization*, 5, pp. 159–90.

Iwai, K. (1984b), 'Schumpeterian dynamics, II: technological progress, firm growth and "economic selection"', *Journal of Economic Behavior and Organization*, 5, pp. 321–51.

Jovanovic, B. and S. Lach (1989), 'Entry, exit, and diffusion with learning by doing', *American Economic Review*, 79, pp. 690–9.

Kaldor, N. and J.A. Mirrlees (1962), 'A new model of economic growth', *Review of Economic Studies*, 29, pp. 174–92.

Kleine, J. (1983), *Investitionsverhalten bei Prozessinnovationen*, Frankfurt/New York: Campus Verlag.

Kwasnicka, H. and W. Kwasnicki (1986), 'Diversity and development: tempo and mode of evolutionary processes', *Technological Forecasting and Social Change*, 30, pp. 223–43.

Kwasnicki, W. (1989), 'Market, Innovation, Competition', discussion paper, Institute of Engineering Cybernetics, Technical University of Wroclaw, Poland.

Levin, R.C., Cohen, W.M. and D.C. Mowery (1985), 'R&D appropriability, opportunity and market structure: new evidence on some Schumpeterian hypotheses', *American Economic Review: Papers and Proceedings*, 75, pp. 20–4.

Lundvall, B.A. (1988), 'Innovation as an interactive process: from user-producer interaction to the national system of innovation', in Dosi, G. *et al.* (eds), *Technical Change and Economic Theory*, London: Pinter.

Malcomson, J.M. (1975), 'Replacement and the rental value of capital equipment subject to obsolesence', *Journal of Economic Theory*, 10, pp. 24–41.

Massé, P. (1962), *Optimal Investment Decisions*, Englewood Cliffs, NJ: Prentice-Hall.

McCloskey, D.N. (1974), *Economic Maturity and Entrepreneurial Decline*, Cambridge: Cambridge University Press.

Nelson, R.R. and S.G. Winter (1982), *An Evolutionary Theory of Economic Change*, Cambridge, Mass./London: Belknap Press of Harvard University.

Noble, D.F. (1984), *Forces of Production: a Social History of Industrial Automation*, New York: Alfred Knopf.

Olmstead, A. (1975), 'The mechanization of reaping and mowing in American agriculture, 1833–1870', *Journal of Economic History*, 33, pp. 327–52.

Rosenberg, N. (1972), 'Factors affecting the diffusion of technology', *Explorations in Economic History*, 10.

Rosenberg, N. (1976), 'On technological expectations', *Economic Journal*, 86, pp. 523–35.

Rosenberg, N. (1982), *Inside the Black Box: Technology and Economics*, Cambridge: Cambridge University Press.

Sahal, D. (1981), *Patterns of Technological Innovation*, New York: Addison Wesley.

Sahal, D. (1985). 'Technology guide-posts and innovation avenues', *Research Policy*, 14, pp. 61–82.

Salter, W. (1960), *Productivity and Technical Change*, Cambridge: Cambridge University Press.

Saxonhouse, G. (1974), 'A tale of Japanese technological diffusion in the Meiji period', *Journal of Economic History*, 34, pp. 149–65.

Schumpeter, J. (1912), *Theorie der wirtschaftlichen Entwicklung*, English translation: *The Theory of Economic Development*, Cambridge, Mass.: Harvard University Press (1934).

Silverberg, G. (1987), 'Technical progress, capital accumulation, and effective demand: a self-organization model', in Batten, D., Casti, J. and B. Johansson (eds), *Economic Evolution and Structural Adjustment*, Berlin/Heidelberg/New York: Springer-Verlag.

Silverberg, G. (1988), 'Modelling economic dynamics and technical change: mathematical approaches to self-organisation and evolution', in G. Dosi *et al.* (eds), *Technical Change and Economic Theory*, London: Pinter.

Silverberg, G. (1989), 'Patterns of evolution and patterns of explanation in economic theory', in Casti, J. and A. Karlqvist (eds), *Newton to Aristotle: Toward a Theory of Models of Living Systems*, Boston/Basel/Berlin: Birkhäuser.

Silverberg, G. (1990), 'Dynamic vintage models with neo-Keynesian features', in OECD, *Technology and Productivity: the Challenge for Economic Policy*, Paris: OECD.

Silverberg, G., Dosi, G. and L. Orsenigo (1988), 'Innovation, diversity and diffusion: a self-organisation model', *Economic Journal*, 98, pp. 1032–54.

Smith, V. (1961), *Investment and Production*, Cambridge, Mass.: Harvard University Press.

Solow, R. (1959), 'Investment and technical progress', in Arrow, K.J., Karlin, S. and P. Suppes (eds), *Mathematical Methods in the Social Sciences*, Stanford: Stanford University Press.

Spence, M. (1984), 'Cost reduction, competition, and industry performance', *Econometrica*, 52, pp. 101–21.

Stoneman, P. (1976), *Technological Diffusion and the Computer Revolution*, Oxford: Clarendon Press.

Terborgh, G. (1949), *Dynamic Equipment Policy*, New York: McGraw-Hill.

von Hippel, E.A. (1988), *The Sources of Innovation*, Oxford: Oxford University Press.

Waddington, C.H. (1976), 'Concluding remarks', in Jantsch, E. and C.H. Waddington (eds), *Evolution and Consciousness*, Reading, Mass.: Addison-Wesley.

Winter, S.G. (1984), 'Schumpeterian competition in alternative technological regimes', *Journal of Economic Behavior and Organization*, 5, pp. 137–58.

# 10. Localized technological change, factor substitution and the productivity slowdown

*Bart Verspagen*

## 1. Introduction

The representation of technological change in a neo-classical framework has recently been subject to severe critique (for example, Dosi, 1988 and the references there). One important issue in most of this criticism is the assumption that technological change is a phenomenon that affects all parts of the production structure to an equal extent. As early as 1969, Atkinson and Stiglitz provided a framework in which this assumption is relaxed. However, their model has not received much attention in theoretical and empirical work on technological change (notable exceptions are Freeman and Soete, 1987; Persico, 1988; and Stiglitz, 1987).

The notion of localized technological change might help to solve the so-called productivity paradox. However, as from most 'solutions' to this paradox, we should not expect a complete explanation from the introduction of localized technological change. This chapter will discuss the possible role of localized technological change in the process of a slowdown of productivity growth. The line of reasoning develops along two main arguments.

First, we will discuss the representation of localized technological change in a neo-classical production function framework. In Section 2, we will propose a generalization of the Atkinson and Stiglitz framework. Second, we will analyse some possible consequences of localized technological change, notably the influence of (sudden) price shocks upon substitution and productivity. Section 3 provides a general discussion of the effects of price shocks. The method of analysis applied in Section 4 is that of computer-simulation. We will present the results of a very simple simulation model in which the effects of price shocks upon productivity are analysed. Section 5 will summarize the main arguments.

## 2. Localized versus global technological progress

### *Production possibilities in neo-classical setting*

The neo-classical production function can be represented as in Figure 1. The curve II', commonly known as the isoquant, gives all (efficient) combinations of factor-inputs C and E, with which it is possible to produce a given amount of output $Q^f$. The isoquant is a set of points which correspond to specific modes of production, or techniques. For example, the point A on II' corresponds to a technique in which a large amount of the factor C and little of factor E is used,

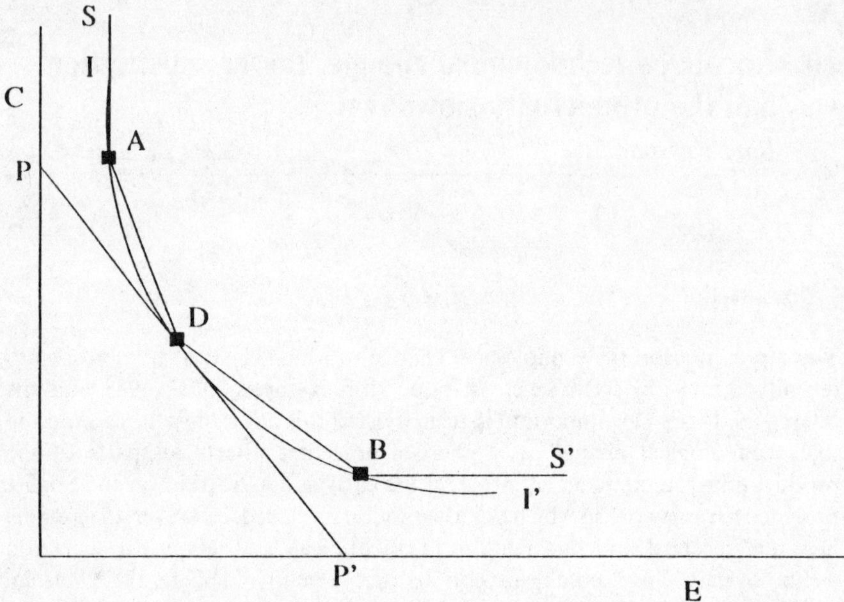


**Figure 1** The production decision in a neo-classical setting

while the point B represents a situation in which a large amount of E, and little of C is used.

The points on the isoquant do not all correspond to single, 'pure' techniques. Some of the points correspond to combinations of two (or more) single techniques, one part of the output being produced by the one technique, and another part by another technique. With a finite number of 'pure' techniques and the assumption of perfect divisibility, the set of production possibilities can be represented as the straight lines connecting adjacent 'techniques', like SS' in Figure 1. In this case, the straight segments correspond to linear combinations of the corner points. In the case of an infinite number of 'pure' techniques, this broken line becomes the smooth isoquant we know so well from the textbook analysis. But even when we do not have an infinite number of 'pure' techniques, we just might approximate the 'real' situation by such a smooth isoquant.[1]

In neo-classical theory, the decision which (combination of) techniques to use is made on the basis of factor prices. There are two ways[2] of dealing with the optimization problem: cost-minimization given a fixed production; and production maximization given a fixed budget. Graphically, these two methods can be explained as follows.

Costs are represented by a so-called iso-cost line like PP' in Figure 1. This line gives all combinations of factors C and E which can be bought with a fixed budget. Cost-minimization given a fixed output corresponds to finding the iso-cost line closest to the origin (i.e. with lowest total costs) with which it is possible to produce the given quantity of output. Output-maximization corresponds to finding the isoquant furthest away from the origin (i.e. the isoquant which yields

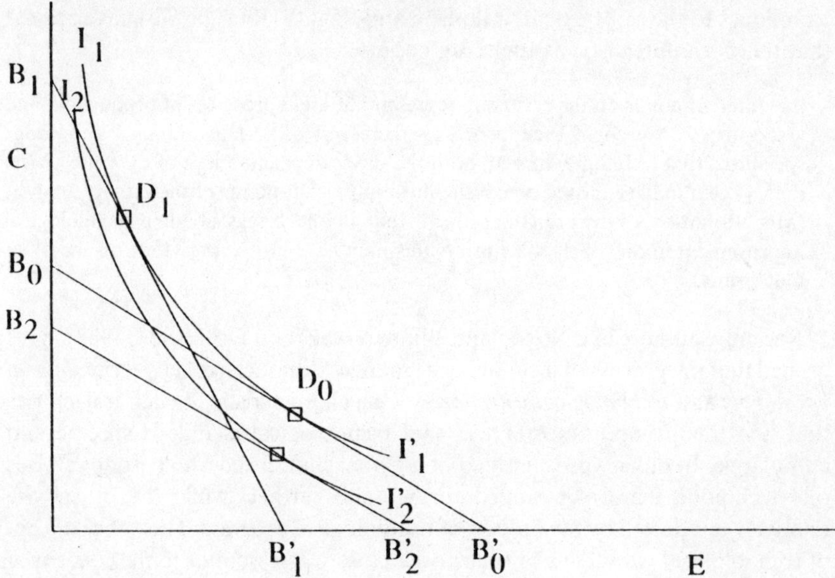


**Figure 2** Substitution and global technological change in a neo-classical setting

highest output) that can still be reached given the iso-cost line. In both cases, the optimal point will be the one in which iso-cost line and isoquant are tangent, as, for example, point D in Figure 1.

In neo-classical theory, changes in the technique used over time can result from two distinct sources. One is *technological progress*, and the other is *substitution*. Figure 2 illustrates these two types of changes. Substitution occurs as a result of changes in factor prices. These changes can be represented as changes in the slope of the iso-cost line. In mathematical terms, substitution occurs because the slope of the iso-cost line changes, and the point of tangency between isoquant and iso-cost line changes accordingly. The iso-cost line $B_1B_1$' in Figure 2 corresponds to a price increase of factor E relative to $B_0B_0$'. The new point of tangency is $D_1$. In $D_1$, less of factor E is used because this factor has become relatively expensive. More of factor C is used because this factor has become relatively cheap. Thus, factor E has been substituted by factor C.

Technological change is represented by a shift of the isoquant towards the origin, as in the case of $I_2I_2$'.[3] Note that the amount of output remains the same in both cases, so that the given output is produced with less inputs in the case of $I_2I_2$'. In the case of cost-minimization, the same amount of output is now produced against lower costs, as represented by the iso-cost line $B_2B_2$'. Thus, technological change is assumed to make *all* techniques more efficient.

## *Localized technological change*

The implicit assumption that is made when technological change is represented by such a global shift of the isoquant is that technological progress affects all techniques to the same extent. Atkinson and Stiglitz (1969, p. 573) have argued that this is an unrealistic assumption, because:

> the different points on the curve still represent different processes of production, and associated with each of these processes there will be certain technical knowledge specific to that technique. Indeed, both the supporters and critics of the neoclassical theory seem to have missed one of the most important points of the activity analysis (Mrs. Robinson's blue-print) approach: that if one brings about a technological improvement in one of the blue-prints this may have little or no effect on the other blue-prints.

Recently, authors like Nelson and Winter (1982) and Dosi (1984, 1988), have argued that the process of technological change is characterized by *technological paradigms* and *technological trajectories*. One characteristic of such trajectories that is of main importance here is that technological change is specific and cumulative. In other words, instead of a global shift of the whole isoquant, it is one technique that is developed further and further, while the others are relatively unaffected by the process of technological progress. Thus, the concept of technological paradigms or trajectories closely corresponds to the concept of localized technological change as introduced by Atkinson and Stiglitz (1969).

Technological change often takes the form of *learning by doing*. This means that the productivity of a technique will increase with the frequency of use, or the amount of output produced with it. Because techniques that are not used are not affected by learning by doing, technological change is localized in this case (Stiglitz, 1987). The same argument holds for R&D expenditures. Firms will not devote efforts to improving techniques they do not plan to use in the (near) future (see Kennedy, 1964 for an early, formal treatment of this argument). This means that the productivity of techniques that are not actually used in the production process will not be affected by technological change, either in the form of learning by doing or in the form of R&D efforts. Therefore, localized technological progress seems a more adequate concept than global technological progress.

Atkinson and Stiglitz (1969) and later on Freeman and Soete (1987) have chosen to represent localized technological change as influencing only one single technique. The first two authors indeed realized that this involved a very heroic assumption: 'In reality we would expect that a given technical advance would give rise to some spillovers and that several techniques would be affected' (Atkinson and Stiglitz, 1969, p. 573; see also Stiglitz, 1987 for a treatment of the role of spillovers). We will label the version of localized technological progress without spillovers as *strong localized technological change*.

In this chapter, we will introduce the less-strong assumption that technological change affects a number off, although not all, production techniques. This means that the efforts of improving one single technique have some spillover effects to related techniques. We interpret the notion of 'related techniques' as those being adjacent on the isoquant. Accordingly, we will label this *weak localized technological change*.

In terms of the isoquants in Figures 1 and 2, we could represent weak localized technological progress by a shift of a part of the isoquant, as in Figure 3. In the

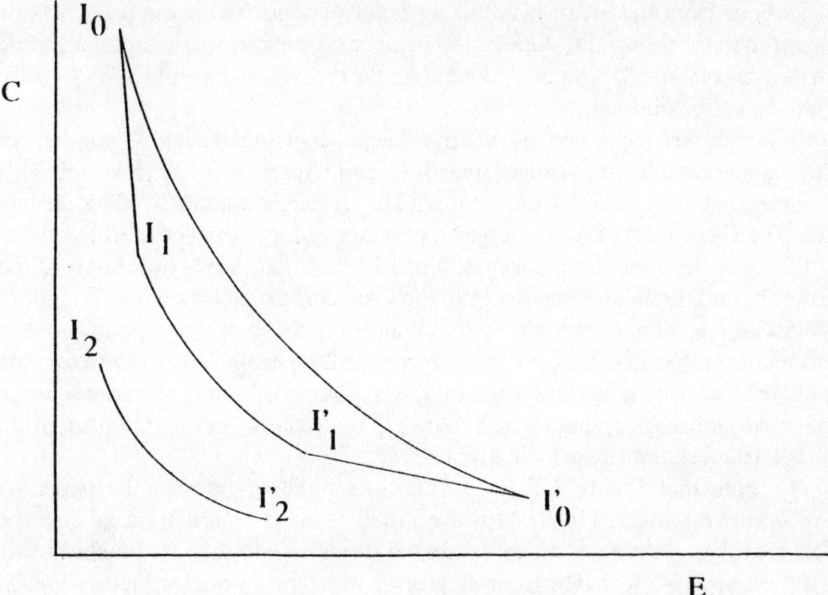


**Figure 3** Weak localized technological change in a neo-classical setting

original situation, it is possible to produce using the range of techniques represented by $I_0I_0$'. Due to technological progress, part of these techniques are improved in the next period, as represented by $I_1I_1$'.

The presence of spillovers might, however, change the form of the shifted part of the isoquant. Because the degree of spillovers may vary among different techniques, it is likely that some points shift more than others. In Figure 3, as in the rest of this chapter, it is assumed that the form of the isoquant does not change. In economic terms, this means that we assume that spillovers are the same for each technique.

However, the points on the shifted segment of the isoquant are not the only ones that are economically relevant for the production decision of the firm. Because it is possible to combine techniques, the straight lines connecting the boundary points of the isoquants $I_0I_0$' and $I_1I_1$' also provide production possibilities that are to be preferred to the points on $I_0I_0$'. This can easily be seen in Figure 3, where the points on the straight lines $I_1$'$I_0$' and $I_0I_1$ are closer to the origin, and therefore correspond to lower costs (higher productivity) than those on $I_0I_0$'. This economic principle also determines the 'cut-off points' of the segment of the isoquant that shifts due to localized technological change. This 'cut-off point' occurs whenever the isoquant would become positively sloped due to very weak spillovers.

When performing the profit-maximization analysis, these straight segments of the isoquant are rather peculiar. If we concentrate on the segment $I_0I_1$, a price ratio which is exactly equal to the (absolute) slope of this segment will yield an indeterminate situation. Each point on the straight segment is as good as its neighbour from the profit-maximizing point of view. When the price ratio is larger than the (absolute) slope of the straight segment, $I_0$ will be optimal, while in the case of a smaller (but still larger than the [absolute] slope of $I_1I_1$' in $I_1$) price ratio $I_1$ will be optimal.

If (localized) technological change keep occurring during a number of subsequent periods, the straight lines $I_1$'$I_0$' and $I_0I_1$ are no longer relevant. This is illustrated in Figure 3 by the isoquant $I_2I_2$', which is characterized by the fact that the E-co-ordinate of the upper boundary point is smaller than the E-co-ordinate of the upper boundary point of $I_0I_0$', and that the C-co-ordinate of the lower boundary point is smaller than the C-co-cordinate of the lower boundary point of $I_0I_0$'. Thus, straight lines connecting the boundary points of the isoquants are positively sloped and economically meaningless. In this case, the only relevant techniques are those on the isoquant $I_2I_2$'. Thus, the consequence of (weak) localized technological change is that relative to the initial situation ($I_0I_0$'), less techniques are available.

We note that Figure 3 looks a little like similar figures in the paper by Atkinson and Stiglitz (1969). However, there is one main difference, being the fact that they restrict (as we noted above) themselves to the case in which only one technique is affected by technological change (strong localized technological change). Therefore, the isoquant in their article that corresponds to our $I_1I_1$' has only straight segments, our smooth segment being replaced by a single point. Their 'isoquant' corresponding to our $I_2I_2$' is nothing more than a rectangle of the Leontief type. This is illustrated in Figure 4, which represents the same process as Figure 3, only for the case of strong localized technological change. Thus, a main argument made by Atkinson and Stiglitz (1969), and also by

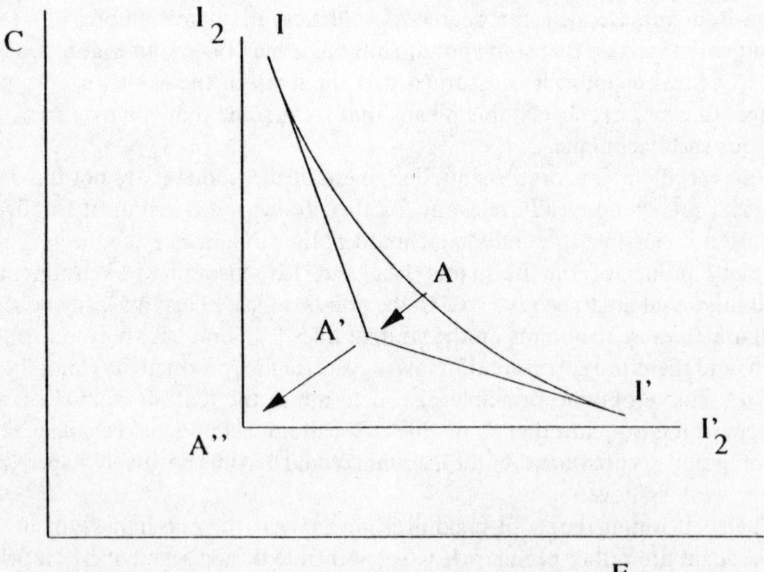


**Figure 4** Strong localized technological change in a neo-classical setting

Freeman and Soete (1987), is that localized technological change diminishes the short-run possibilities for substitution. With strong localized technological change, constant improvements of one single production technique lead to a Leontief shape.

In the case of strong localized technological change, one would have to conclude that the smooth convex isoquant so familiar from neo-classical theory, is something that can only exist in the human mind but not in practice. This is so because even if such an isoquant exists in the first place, it will eventually be replaced by a Leontief production function. This again means that the smooth convex isoquant can only have existed at one point in time, after which the process of localized technological change transformed it, so that it could never take back its old shape again. In the case of weak localized technological change, however, the smooth convex isoquant remains a vital concept, even if it is only valid for a certain range of production techniques.

In the model we present here, the substitution possibilities are limited too. However, this limitation does not take the extreme form of the Leontief rectangular isoquants, because we assumed that substitution is still possible among the points on the shifting segment of the isoquant. We will investigate the possibilities for short-run and long-run substitution in more detail below.

## 3. Price shocks and the substitution effect

When technological change is localized, the firm should take into consideration future factor prices when deciding which technique to use. To see this, imagine that the firm knows that after a certain number of periods the factor price ratio will decrease drastically and will remain constant thereafter. In this case, it should consider switching to the technique corresponding to that new factor price ratio before the price shock occurs. This would improve productivity in the period after the price shock.

But what happens if such a price shock is not foreseen? In order to answer this question, we look at Figure 3 again. Since the range of $I_2I_2$' in Figure 3 is relatively small, substitution on this isoquant is rather limited. As long as relative factor price changes are relatively small (i.e. small changes in the slope of the iso-cost line), substitution will take place as in the normal case of a full-range isoquant. However, if factor prices change more drastically, there will no longer be a point of tangency between $I_1I_1$' and the iso-cost line.

This case is illustrated in Figure 5. $B_0B_0$' is the iso-cost line before the price shock, while $B_1B_1$' is the iso-cost line after the price shock. The solid part of the curve $I_1I_1$' is the isoquant that has resulted from a continuous process of technological change. When technological change would have been global, the whole range of the isoquant (including the dotted part) would have been valid. When the input price ratio changes drastically, as, for example, from $B_0B_0$' to $B_1B_1$', there is no longer an 'optimal' solution on the localized isoquant. In Figure 5, E* would be the solution to the profit-maximizing exercise in the case of global technological change. But because E* is on the dotted part of the isoquant, it cannot be the solution in case of localized technological change. In

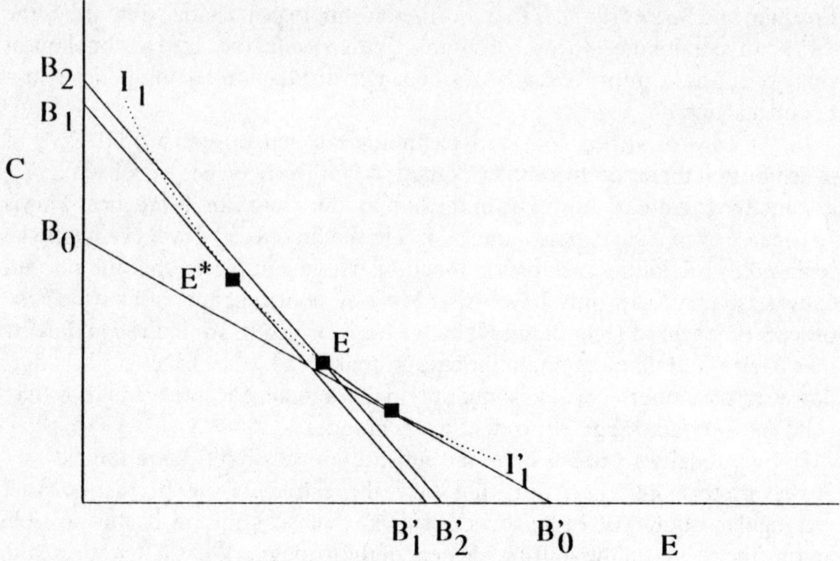


**Figure 5** The effect of a factor price shock in case of localized technological change

the latter case, E will be the (corner-) solution.

It is obvious that the corner-solution on the solid part of $I_1I_1$' is 'inefficient' relative to E*. Imagine that the firm could have anticipated the price shock, which is assumed to last. In that case it certainly would have considered developing and applying a technique more adequate for the expected price ratio (see also Atkinson and Stiglitz, 1969, pp. 574–5; and Stiglitz, 1987, p. 131). However, since the price shock was not foreseen, the firm is now caught in a situation in which it is forced to use an inefficient technique.

In the long run, the firm can devote its innovation (R&D) efforts to finding a new set of techniques better suited for the new price ratio. But if technological change takes place by a process of learning by doing, then applying a new technique that has not been used before will most certainly involve initial productivity losses. In order to arrive at the level of productivity of the technique currently in use, a new process of R&D and of learning by doing must be started.

In order to undertake this techniques-switch, the firm, of course, must be convinced that the price change will last. Moreover, it will have to be persuaded that the rewards for developing a new technique are so high that it is profitable to switch to the new technique. It could be the case that the productivity differences between the technique currently used and the new technique are so big that the firm will keep on using the old technique, at least for a short period of time.

We can ask the question whether the firm will switch production techniques at all. Would it not be possible that the productivity gap between the old and the new technique is so big that the benefits of having a more suitable factor ratio do not compensate the (initial) productivity losses? In order to answer this question,[4] we consider a simple production function of the form

$$Q_t = T_i K_t^{\alpha} L_t^{1-\alpha} \quad (1)$$

where the subscript $t$ denotes time, Q is production, and K and L are production factors. We assume that the firm has a choice between two different 'trajectories' of localized technological change. The factor representing technological change is $T_i$, which is equal to $t+1$ minus the introduction time of the trajectory. Each of these trajectories $i$ has a fixed average factor ratio[5] denoted by $k_i$, which determines the (expected) production costs of a trajectory in each period. Dropping time subscripts, we can define the factor ratio by

$$k_i = \frac{K_i}{L_i} \quad (2)$$

Substituting definition (2) in the production function (1), we can obtain a cost function $c_i(t)$.

$$c_i(t) = \frac{\frac{wQ}{k_i^{\alpha}} + \frac{rQ}{k_i^{\alpha-1}}}{T_i} = T_i^{-1}\beta_i \quad (3)$$

$$\text{with } \beta_i = \frac{wQ}{k_i^{\alpha}} + \frac{rQ}{k_i^{\alpha-1}}$$

where $w$ and $r$ denote the exogenous prices of L and K, respectively. The cost function describes the production costs of trajectory $i$ at time $t$. Note that it is not a cost function as we normally formulate it. A 'normal' cost function gives the cost corresponding to efficient combinations of production factors, resulting from different factor prices. The cost functions presented here can be inefficient as a whole. The βs reflect the suitability of the factor price ratio to the factor ratio. When the combination of these two is optimal (i.e. when the trajectory's $k$ corresponds to the $k$ that would result from optimizing the 'standard' cost function), the β will be minimal.

Because technological change is localized, the firm has to consider not only current production costs but also future costs. Thus, for the decision about which technique to use, the firm has to consider the cumulative costs of producing with technique $i$.[6] With expectations of fixed factor prices (i.e. fixed βs) and fixed sales, the cumulative production costs $C_i(t)$ are defined[7] by

$$C_i(t) = \int c_i(t)dt = \int (\beta_i T_i^{-1})dt = \beta_i \ln T_i \quad (4)$$

For a given factor price ratio and equal $T_i$, the two cumulative cost functions intersect only at the point where $T_i = 0$. For $T_i > 0$, one cumulative cost function will always be higher than the other. This is depicted in Figure 6, where the functions A and B correspond to different trajectories $i$.[8] The slope of the curves in Figure 6 diminishes with time. This is caused by the effect of productivity growth, due to (localized) technological change. Thus, at $t_0$, when the firm faces the initial choice which trajectory to choose on the basis of the expectation that relative factor prices will not change, it will choose the one with the lowest $\beta_i$, i.e. the one with the factor ratio $k_i$ best suitable to the factor price ratio.

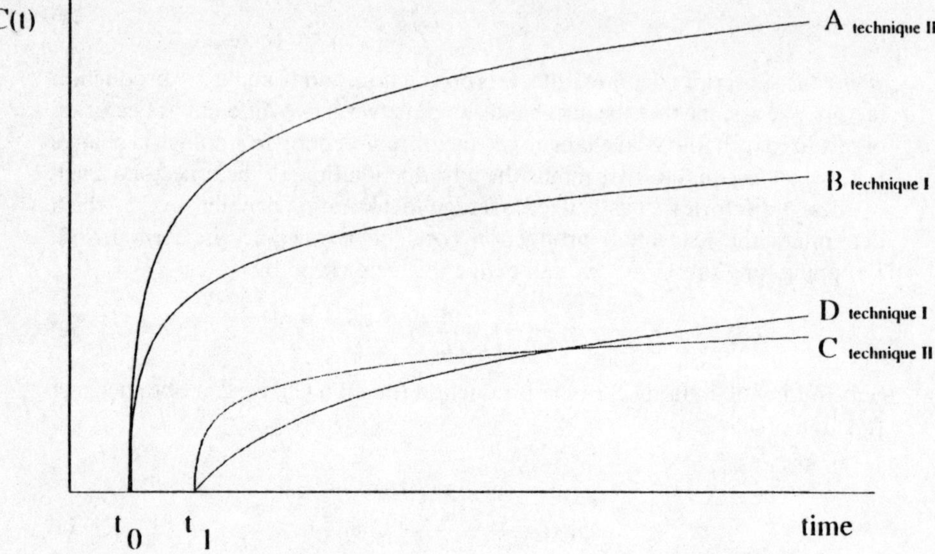


**Figure 6** The decision of switching techniques after a price shock at period $t_1$

Imagine now that after a while (at $t_1$) an unexpected factor price shock occurs, which reverses the sequence of βs. Thus, the trajectory with the lower β now becomes the trajectory with the higher β and vice versa. The decision which technique to use from the moment the price shock occurred is a little bit more complicated now, because the state of development of the two trajectories is no longer equal. The firm has to choose between developing the previously unused technique and continuing with the 'old' technique. The former alternative means that there will be an initial productivity offset, i.e. that $T_i$ 'starts' at 1 again. The latter alternative means that $T_i$ just continues to rise from its current level, but that the factor ratio is not suitable. Graphically this decision can be represented by the curves C and D in Figure 6. The curve D corresponds to the cumulative costs associated with a continuation of use of the old technique. Curve C corresponds to the introduction of the other trajectory. In the neighbourhood of $t_1$, the curve C has a steeper slope than curve D, and is thus above curve D. But at a certain point, the two curves intersect, and the curve C lies below D beyond that point.

This means that although the cumulative costs of introducing the new trajectory are initially above those of continuing with the old trajectory (due to a productivity offset), at a certain point in time the total costs of introducing the trajectory will be lower. Now the decision whether or not to apply the new trajectory depends on whether this 'certain point in time' lies within the decision horizon of the firm. If it does, the firm will switch, otherwise (for example, because the firm cannot form expectations about events happening after a certain time delay) it will stick to the old trajectory. In the rest of the chapter, we will assume that the decision horizon of the firm is infinite, so that it will switch for certain.

Although the above analysis shows that it might be profitable to switch to the new technique after a price shock, it has not been shown that it is necessary to apply the technique immediately after the shock occurs, i.e. at point $t_1$ in Figure 6. The analysis of Figure 6 leaves open the possibility that the firm keeps applying the old technique for some time, and develops the new technique in the R&D laboratory. Only when the new technique has reached the productivity level of the old technique (the cutting point of C and D) will it then be applied. This would, of course, reduce the productivity offset effect to a considerable extent. However, if we assume that *learning by doing* is an important part of technological progress, it will still be necessary to switch techniques before the cutting point is established in the R&D laboratory.

The process of structural change at the firm level will most certainly have some major macroeconomic effects. The productivity offset that the individual firms face if they switch techniques will manifest itself at the macro level as a productivity slowdown. We can identify a few key variables which are likely to influence the exact shape of this process of structural change. Without wanting to suggest these are the only important variables, we will focus in the rest of this chapter on two of them. These are the average adjustment speed of the firm to the new price ratio and the length of the period in which localized technological change has affected the productivity of the techniques currently in use.

The speed of adjustment to the new price ratio will most likely have a macroeconomic effect. Intuitively, one would expect that the higher the speed of adjustment is, the more acute and pronounced the resulting productivity slowdown will be. If firms adapt their techniques slowly, and only a few of them switch at the same time, the macroeconomic effects in each separate period will be relatively small but the adjustment process will take a long time. When, on the other hand, the speed of adjustment is high, the effects in each separate period will be large, but the process will be over sooner.

The length of the period in which localized technological change has affected the techniques in use before the price shock will, of course, be related to the size of the productivity advantage over the techniques unaffected by technological change. This will affect the size of the productivity offset the firm (and the economy as a whole) experiences after the new technique is adopted.

In the next section we will introduce a simple computer simulation model through which we will investigate the macroeconomic effects of the price shocks.

## 4. A simulation of the macroeconomic consequences of a price shock

In this section, we will investigate the macroeconomic consequences of the microeconomic effects of price shocks analysed in the previous section. We will assume that the technological level of individual firms differs. It is suggested that a productivity slowdown as most industrial economies experienced in the early 1970s can (partly) be due to the effects of localized technological change in combination with sudden price shocks.

*The model*

The computer simulation model has the following characteristics. There are 100 firms in the economy, which produce one homogeneous commodity. The market is assumed to be (geographically) divided: each firm has a fixed part of the market equal to the market share of each of the other firms. Total production is constant. Prices of inputs and outputs are exogenous to the firm. Production takes place according to a Cobb Douglas production function, with two inputs called energy and composite input.[9]

Innovation is assumed to take the form of process innovation, and is of the (Hicks) neutral kind. Firms do not all use the same technology, but are initially distributed randomly over different states of technological knowledge (i.e. they all face a different value of the technological shift factor). This is done to simulate the process of diffusion of technology. As a consequence, each firm has its own specific level of productivity. Each period, the isoquant of all firms shifts by a given rate of technological progress. Technological change is, however, not global, but localized. The isoquants corresponding to the technology the firm uses are valid only for a certain range of values of energy/composite input (compare Figures 3 and 4 above). The boundaries for which the recent isoquants are relevant are different for each value of the technological shift factor in the production function and are defined as the range between the optimal points for the boundaries between which the factor price ratio is assumed to fluctuate in normal situations.

Each period, each firm determines the optimal quantities of energy and composite input to use, given the fixed output, the technology it uses, and the factor price ratio. These individual inputs are then aggregated and the macroeconomic productivity indexes (energy productivity, composite input productivity, and total factor productivity) are determined. At a certain point in time, the factor price ratio is assumed to change to a value outside its 'normal' range. Initially, this will force firms to produce at a corner-solution of their isoquant. However, in the long run, firms will switch to other techniques, more suitable to the new factor price ratio.

Thus, immediately after the factor price shock, firms will start adjusting their R&D efforts towards trying to find a set of techniques better suited to the new price ratio. It is assumed that for the current purposes this search process (see Nelson and Winter, 1982) can be represented by a stochastic model. Therefore, it is not likely that all firms succeed in their efforts at the same time. The adjustment process is modelled here as a simple random chance that the firm will switch to a new production technique, using the optimal input bundle for the new price ratio. Associated with this new technique is a lower level of productivity (or more formally a smaller value of the technology shift factor in the production function).

The adjustment speed to the new technique is altered through varying the (average) number of firms that will shift to the new technique in one period. Formally, this is done by varying the chance that a switch will occur. The other simulation variant that will be discussed is the length of the period of localized technical change before the price shock occurs. This is of course simply set by varying the period in which the price shock occurs.

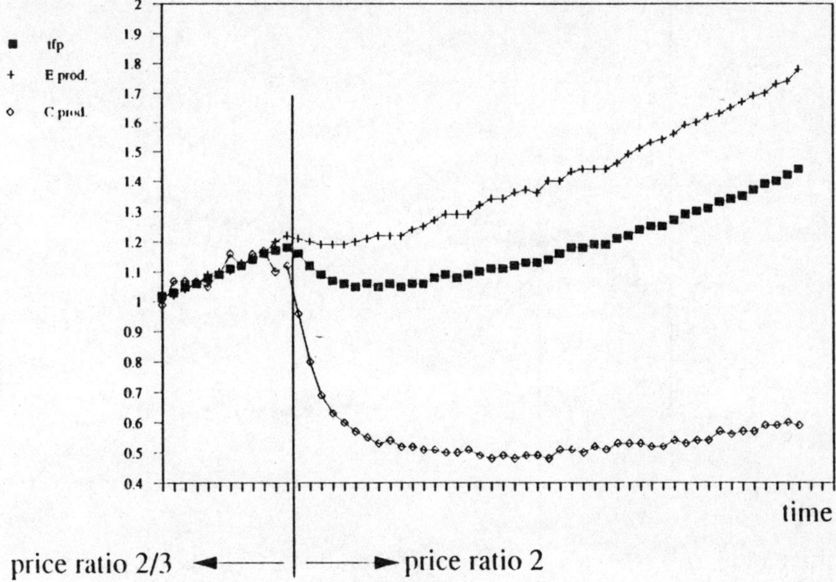


**Figure 7a** Simulation run with intermediate adjustment speed to new price ratio

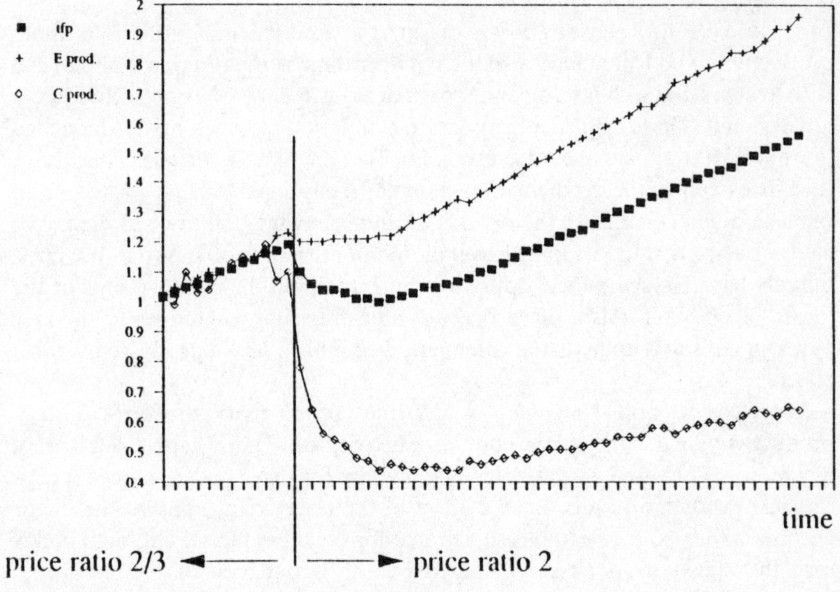


**Figure 7b** Simulation run with high speed of adjustment to new price ratio

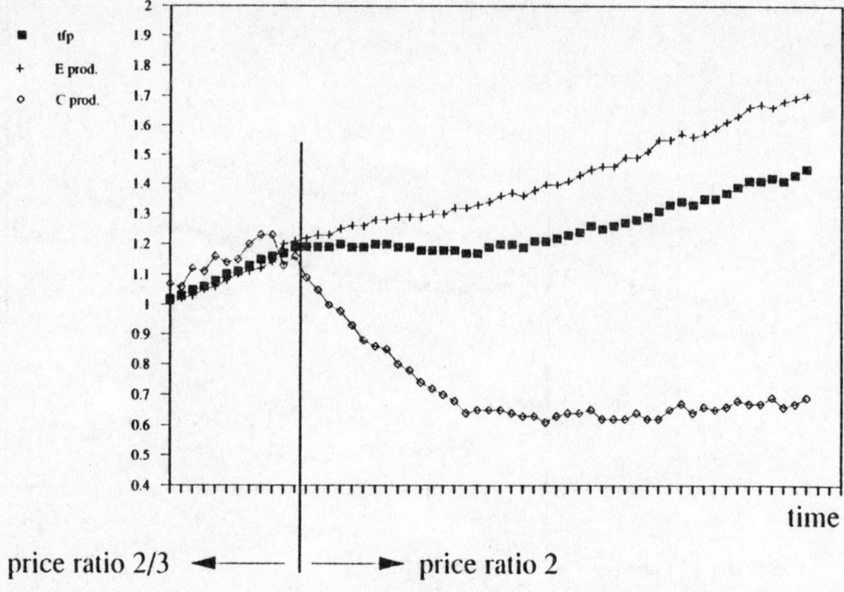


**Figure 7c** Simulation run with low speed of adjustment to new price ratio

*The results*

*(a) Varying the adjustment speed*

In Figure 7, the growth paths of energy productivity, composite input productivity and total factor productivity for three runs — one with high, one with low and one with intermediate speed of adjustment to the new technique — are depicted. These results correspond to a situation in which prices fluctuated 'normally' (i.e. an average price ratio of 2/3) in the first 12 periods. This means that firms can optimize their input bundle 'freely', maximizing profits in an unconstrained context. In the periods 13 and following, the price of the factor energy is tripled, thus yielding an average factor price ratio of 2. As a result, firms initially have to locate their optimal input bundle on the corner-points of the localized isoquant. Also, the process of adjustment of techniques to the new factor prices starts immediately after period 12. This causes a productivity offset effect.

We have to note that the two 'partial' productivity measures, energy productivity and composite input productivity in Figure 7 give a bit of a troublesome picture of what is really going on. Since we have assumed a neo-classical production function, the effect of the price change is that the factor becomes more expensive and is substituted for the other factor, although in this case the substitution effect is limited by the boundaries of the 'localized' isoquant. This substitution effect has, of course, an immediate effect on productivity, stimulating productivity of the factor being substituted and lowering the productivity of the other. The three figures show that the trends for energy productivity clearly overstate the actual productivity growth. Indeed, this trend does hardly show a decline after the price shock. This is, of course, due to the substitution effect, which offsets the decline in productivity associated with the technological switch. The series for composite input productivity, on the other hand, shows a marked decline, which overstates the real productivity trend because of the negative substitution effect associated with the price shock.

Therefore, we should base our conclusions mainly on the trend in total factor productivity. As Figure 7 indicates, the latter is nothing more than a weighted average of the two partial productivity indexes, using the factor shares in total costs as weights.[10] In Figure 7a, the trend for total factor productivity shows a negative growth for some time after the price shock, indicating that firms are switching production techniques. After a period of negative productivity growth, total factor productivity takes up again. In the beginning the upswing is slow, but after a while the strong upward trend of the period before the price shock is restored. This point corresponds to the point in time where the majority of firms have switched techniques, and the 'long run' equilibrium is again restored.

It seems plausible that the precise trend in total factor productivity depends on the speed of adjustment to the new price ratio. Therefore, we present the results of two more simulation runs, in which the adjustment speed is changed. The results for the higher adjustment speed are presented in Figure 7b. As could be expected, the productivity offset is somewhat more pronounced, but the turning-point comes earlier. In the case of the lower adjustment speed (Figure 7c), the offset is less pronounced (indeed there is almost no negative growth of total factor productivity at all), but the productivity slowdown phase lasts longer in this case.

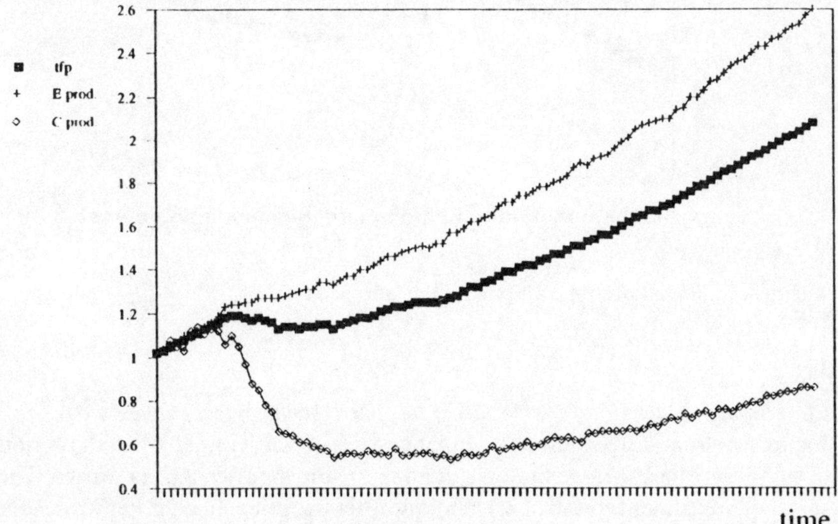


**Figure 8a** Simulation run with price shock after period 12

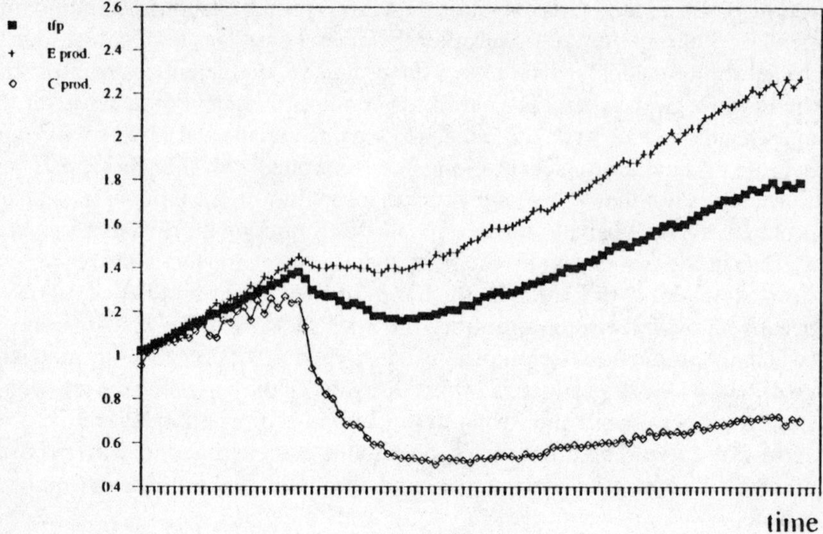


**Figure 8b** Simulation run with price shock after period 24

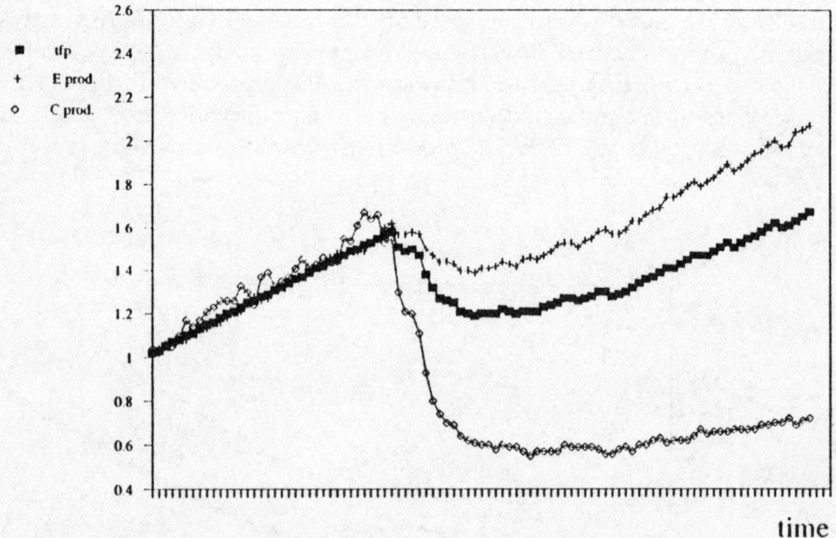


**Figure 8c** Simulation run with price shock after period 36

*(b) Varying the length of the period of localized technological change*

Figure 8 presents three different simulation runs in which the length of the period of cumulative localized technological change before the price shock is varied. The speed of adjustment to the new technique after the price shock is kept constant and is a little bit higher than the speed in the simulation run in Figure 7c.

The lowest point in the time series for productivity (tfp) is about equal for all three variants, but the level from which productivity falls after the price shock is, of course, lower the sooner the price shock occurs. Since the length of the period needed to reach the upturn of the productivity series is also (about) equal for the three variants, the level of productivity reached after a certain time is different among the three variants. In the first variant, the one with a relatively short period of cumulative localized technological change before the price shock occurs, productivity at the end of the simulation period is highest. In the third variant, the one with the longest period before the price shock, the productivity level at the end of the simulation period is lowest.

## 5. Summary and conclusions

We have argued that technological change should be represented by localized shifts of the neo-classical production function rather than by global shifts. Localized technological change can be defined in a strong sense, influencing only one single production technique, and in a weak sense, influencing a (limited) range of techniques. Atkinson and Stiglitz found that localized technological change in a strong sense diminishes the short-run possibilities for substitution between factor inputs to zero. We found that localized technological change in a weak sense reduces short-run possibilities for substitution too, but still leaves possibilities for substitution within certain input price limits.

A straightforward graphical analysis shows that in this case sudden shocks of input prices can bring the individual firm out of its long-run equilibrium position. We argued that the 'induced technological switches' to sets of techniques more suited to the new factor price ratio would yield periods of macroeconomic productivity slowdown. In a simple production function framework with a sufficiently large decision horizon for the firm, it can be shown that these 'induced technological switches' are, indeed, rational from a profit maximizing point of view. These switches cause a productivity slowdown.

A computer simulation of the macroeconomic consequences of this slowdown effect shows that the speed of adjustment to new technological areas and the length of the period of cumulative localized technological change before the price shock occurs are important determinants of the precise productivity path. A high adjustment speed yields relatively short, but violent, periods of productivity offset. A low adjustment speed yields long and calm periods of productivity decline. The longer the period of cumulative localized technological change before the price shock occurs, the stronger is the productivity offset, and the lower levels of productivity after a fixed period are.

In the computer simulation experiment we used two inputs, named energy and composite input. This was, of course, done to suggest the link between the model and the actual historical trend in the 1970s and early 1980s. In this period, the world economy experienced two major oil price shocks, and at the same time showed a marked decline in productivity. The model presented here suggests that the oil price shock, through inducing a process of technological switches to areas outside the techniques normally in use, is one of the determinants of the observed productivity development.

At the same time, however, we have to realize that this argument makes only limited sense. As we know from the statistics, the productivity slowdown set in already *before* the first oil price shock, and accelerated after 1973 (see, for example, Link, 1986, p. 30). Thus, one can only argue that a switch of techniques due to the oil price shock was a factor that added considerably to the productivity slowdown effect set in motion by another factor. With regard to further considerations about the empirical value of the arguments made above, only further empirical research can give more definite answers.

**Acknowledgements**

The author thanks Charles Cooper, Giovanni Dosi, Theon van Dijk, Rene Kemp, Bengt Åke Lundvall, Huub Meijers and Rombout de Wit for helpful comments on earlier versions of this chapter.

**Notes**

1. Obviously, the approximation becomes better the larger the number of 'pure' techniques there are. For a more detailed description of the parametrization of the production process, see Varian (1984).
2. The fact that these two approaches lead to the same outcome is discussed in the literature on 'duality' of the production optimization problem.
3. In more detailed analyses, a distinction is made between neutral, labour-saving and capital-saving technological progress. These concepts can be defined in different ways (i.e. as defined by Harrod, Hicks and Solow). Here, as in most of the literature, we will only take into account the case of (Hicks) neutral technological change (see also the chapter in this book by de Wit).
4. Stiglitz (1987, p. 147–9) asks the question which technique the firm should choose, given that none of the available techniques has been developed and that future factor prices are more or less known, but the 'learning functions' of techniques vary. Although this question is, of course, related to the one that is investigated here, it does not shed any light upon the effect of sudden price shocks. The analysis he presents leads to the conclusion that firms should take into account future costs instead of present costs alone. This is, of course, in agreement with the conclusion reached below.
5. This average factor ratio corresponds to the middle-point of the segment of the isoquant that shifts in the case of *weak* localized technological change.
6. We will assume a 0 rate of interest in the implicit calculations of present value below.
7. In formula (4), we set the integration constant to 0.
8. The results of the graphical analysis of the cumulative cost functions presented here can, of course, also be obtained by the corresponding procedure of solving the logarithmic equations for time *t*.
9. The production function has the following functional form: $Q_t = A_t E_t^{0.25} C_t^{0.75}$
10. See for example Solow (1957) and Englander and Mittelstädt (1988) for a treatment of the calculation and interpretation of total factor productivity.

## References

Atkinson, A. and J.E. Stiglitz (1969), 'A new view of technological change', *Economic Journal*, 79, pp. 573–8.

Dosi, G. (1984), *Technical Change and Industrial Transformation*, London: Macmillan.

Dosi, G. (1988), 'Sources, procedures and microeconomic effects of innovation', *Journal of Economic Literature*, 26, pp. 1120–71.

Englander, A.S. and A. Mittelstädt (1988), 'Total factor productivity: macro-economic and structural aspects of the slowdown', *OECD Economic Studies*, pp. 17–56.

Freeman, C. and L. Soete (1987), 'Factor substitution and technical change', in Freeman, C. and L. Soete (eds), *Technical Change and Full Employment*, London: Basil Blackwell, pp. 36–48.

Kennedy, C. (1964), 'Induced bias in innovation and the theory of distribution', *Economic Journal*, LXXIV, pp. 541–7.

Link, A.N. (1986), *Technological Change and Productivity Growth*, Harwood Academic Publishers.

Nelson, R.R. (1981), 'Research on productivity growth and productivity differences: dead ends and new departures', *Journal of Economic Literature*, 19, pp. 1029–64.

Nelson, R.R. and S.G. Winter (1982), *An Evolutionary Theory of Economic Change*, Cambridge Mass.: Harvard University Press.

Persico, P. (1988), 'R&D expenditure and localized technological progress: a theoretical approach to explain some recent R&D performances in manufacturing industry', *Studi Economici*, 34, pp. 13–23.

Solow, R.M. (1957), 'Technical progress and the aggregate production function', *Review of Economics and Statistics*, 39, pp. 312–20.

Stiglitz, J.E. (1987), 'Learning to learn, localized learning and technological progress', in Dasgupta, P. and P. Stoneman (1987), *Economic Policy and Technological Performance*, Cambridge: Cambridge University Press, pp. 123–53.

Varian, H.R. (1984), *Microeconomic Analysis*, second edition, New York/London: Norton.

# 11. Vintage capital and R&D based technological progress

*Adriaan van Zon*

## 1. Introduction

The purpose of this chapter is to focus attention on the possible influence of R&D based technological progress on changes in the size and composition of the capital stock, and the implications of the latter for changes in productivity. In order to do so, we describe a production framework in which labour and capital, given the impact of R&D based technological progress, are combined in the most profitable way. As a starting-point for this exercise we use the notion that capital is not homogeneous and that productivity increases due to R&D efforts are (partly) embodied in different vintages of machinery and equipment.

The vintage framework chosen is of the putty-clay type. This means that producers may choose from a wide range of ex-ante labour intensities of production, whereas once they have made their choice, both labour requirements and capital requirements per unit of capacity output are more or less fixed. This approach contrasts with the aggregate production function approach in the sense that substitution between labour and capital is a phenomenon which is directly connected with the installation of new machinery and equipment. Another form of substitution is, however, the replacement of old machinery and equipment with new equipment which is more profitable at the ruling wages and prices. In this sense, substitution is a truly marginal phenomenon: it is the result of both the decision on the composition of the newest vintage in terms of initial labour and capital coefficients, and the decision on the scale of production on the newest vintage as influenced by the economic scrapping of existing vintages.

The outline of this chapter is as follows. In Section 2 we will pay some attention to the 'measurement' of technological progress based on growth accounting. We describe the assumptions underlying this approach, and focus our attention on one of these in particular, i.e. the homogeneity of capital. In Section 3 we describe the outlines of a putty-clay vintage model as an alternative to the neo-classical aggregate production function approach. We also try to link the different types of technological progress to R&D efforts. The chapter ends with a summary and some concluding remarks.

## 2. Technological progress and factors of production

### *2.1 Total factor productivity*

Using the concept of a production function, the development over time of output can be written as a function of time itself and of the physical factors of production, which are usually taken to be labour and capital. Indeed, as Solow pointed out,[1] the growth of output can be taken to consist of two parts: a change in the volume and intensity of use of the physical inputs; and a shift over time of the production function itself. It is this latter shift which is associated with technological progress.

In order to obtain a quantitative picture of the relative importance of the 'non-physical' factors of production in the explanation of the rate of growth of output, growth accounting exercises usually assume a Cobb-Douglas form for the production function. In the process, a number of simplifying assumptions are made.[2] Perfect competition is assumed to prevail on the various markets, with labour and capital being paid their marginal product and entrepreneurs acting as profit maximisers. Changes in relative prices are presumed to lead to instantaneous and costless adjustments in the capital/labour ratios. Economies as well as diseconomies of scale are assumed not to exist or to be irrelevant. In addition, it is assumed that capital as well as labour are homogeneous. With respect to capital this means that a unit of capital installed in year *t* is the same as a unit installed earlier/later from the point of view of its productivity.

Taken together, these assumptions imply that the impact of technical progress on the rate of growth of output can be proxied by the difference between the rate of growth of output and a weighted sum of the rate of growth of the labour and capital inputs. Let us assume that in its general form the production function can be written as:

$$Q = F(t,L,K) \tag{2.1.1}$$

where Q is output, t is time, L is labour input and K capital input.

F is assumed to be a linearly homogeneous and twice differentiable function with respect to L and K. Assuming a constant rate of technical progress equal to g, and the share of labour in output to be equal to $\alpha$, the rate of growth of output can be written as:

$$\dot{q} = g + (1-\alpha)\dot{k} + \alpha\dot{l} \tag{2.1.2}$$

where:

$$\dot{q} = \partial \ln Q/\partial t$$
$$\dot{k} = \partial \ln K/\partial t$$
$$\dot{l} = \partial \ln L/\partial t$$

From equation (2.1.2) it follows that the rate of growth of output is determined by the rate of technical progress itself and by the rate of growth of the weighted capital-labour combination. Alternatively, relaxing the assumption of a constant rate of technical progress and having measures of output and both labour input and capital input in hand, the equation can be used to calculate the rate of technical progress (or 'total factor productivity growth') as the difference between the rate of growth of output itself and the rate of growth of the weighted labour-capital combination. It is this approach which has often been followed in order to obtain measures of the impact of technical progress on output growth.

In the framework of equation (2.1.2) measurements of total factor productivity are usually related to the progress of time and/or to R&D efforts. With respect to the latter variable, one usually distinguishes between several types of R&D, i.e. basic versus applied, public versus private and process versus product. R&D is normally incorporated into the production function by the addition of a new independent variable. In the common Cobb-Douglas formalism output is typically written as:[3]

$$Q = A e^{\lambda t} R^{\beta} K^{1-\alpha} L^{\alpha}$$

where R represents the R&D capital accumulated by the firm or industry and is sometimes separated into basic research capital and industrial R&D capital. There are several difficulties with this approach. To begin with, if R is zero there is no output. This is unreasonable. Secondly, introducing R as a new independent variable excludes the possibility of modelling the links between R&D effort, scrapping behaviour and capital and labout growth.[4] In order to model this linkage, a vintage modelling framework is adopted.

## *2.2 Heterogeneous capital*

The homogeneity assumption of capital together with the other assumptions mentioned above lead to useful simplications but do so at the expense of plausibility and accuracy. During the last decades vintage models of production have been developed, and it is the vintage composition of the stock of capital which provides the means to be more explicit about the impact of technical progress on the growth of output. The basic feature of vintage models lies in the notion that the quality of new capital changes (improves) over time, reflecting increases in technical knowledge and their subsequent embodiment in investment goods. In this sense capital is heterogeneous:[5] different vintages cannot simply be combined into an aggregate production function.

At the heart of the vintage framework is a model of how entrepreneurs decide how much to invest in each vintage. This decision involves two elements. Based on their expectations of future prices for capital and labour, and of demand and the rate of technical progress (modelled here by growth in R&D stocks), producers must choose an investment from among the available types of machinery which gives them an optimum capital-labour ratio. In addition, they must decide how much of previous vintages to scrap in favour of the latest vintage. These two choices are interlinked and must be solved simultaneously.

To model the first choice, this chapter adopts a 'putty-clay' vintage model[6] based on:

- an ex-ante part: each year, a range of investment goods characterised by different capital-labour ratios is available to producers;
- an ex-post part: once producers choose a particular type of investment good (i.e. a particular capital-labour ratio), this remains fixed except for changes in working hours, machine hours and the influence of disembodied technical progress.

Thus, once producers have chosen a certain capital-labour ratio from amongst the available investment goods they cannot alter it very much. It is this relative ex-post immalleability of vintage capital which calls for an ex-ante decision rule that includes expectations of future wage costs: i.e. it must take account of the fact that producers have to face the consequences of their choice for a considerable period of time. It is clear that such a decision rule is less relevant in the case of 'putty-putty' capital which can be transformed without any cost so as to be optimum at all relative prices for capital and labour; and 'clay-clay' capital where there is just one type of new capital available. Of the three, the 'putty-clay' models are the most realistic.[7]

To model the second choice regarding scrapping behaviour, the paper assumes that entrepreneurs attempt to maximise their expected profits over all vintages. Old vintages are scrapped when more profit could be earned by switching the corresponding production to the latest vintage. This is a stronger condition than merely requiring that vintages cover their marginal costs: producers will scrap those vintages which generate a rent (or quasi-rent) per unit of output which is lower than the rent on the newest vintage per unit of output (corrected for the associated scrapping costs per unit of output). We will come back to this later.

Producers link the two choices together, so that their initial choice of the capital-labour ratio on the newest vintage (the 'putty-clay' part of the model) will be in line with their scrapping decision, i.e. both types of decisions are based on the assumption that producers are actively trying to maximise profits both at the level of the new vintage and at the level of the aggregate stock of capital. Rather than assuming perfect foresight[8] the chapter uses a simple adaptive expectations model in order to describe the expected development over time of the relevant economic decision variables.

## 3. The vintage model

### *3.1 Expectations*

As was mentioned above, we use a simple adjustment scheme in order to derive the expected values of the relevant economic variables which control the ex-ante choice with regard to the initial labour intensity of production on the newest vintage, i.e. the wage rate (w), the price index of output (P) and the index of working hours $h$. We will assume that the expected rate of growth of the variable in question is the weighted average of the observed rates of growth of the variables in question over the last $q$ years. Furthermore we assume that the expected levels of the variables in question for time $t'$ are generated from the observed levels at time t and the expected rates of growth for the transition from t to $t'$. Denoting expected values by a superscript $e$ and the rate of growth of variable $x$ from t–1 to t by $\dot{x}_t$, the above can be summarised as follows:

$$\dot{x}_t^e = \sum_{i=t-q+1}^{t} s_i \dot{x}_i \tag{3.1.1}$$

The variables $s_i$ are predetermined weights which add up to unity. Using equation (3.1.1) expected levels can be written:

$$x^e_{t,t'} = x_t\,(1 + \dot{x}^e_t)^{t'-t} \tag{3.1.2}$$

where $x^e_{t,t'}$ is the expectation of variable $x$ for time $t'$ at time t.

### 3.2 Choosing an optimum capital intensity of production

As was mentioned in paragraph 2.2, we will use a putty-clay vintage model in order to describe the production structure of some sector of industry. At this stage we will not dwell upon the linking of technical progress to R&D efforts. Instead we use simple functions of time in order to describe the various types of technical progress, i.e. embodied and disembodied technical progress. We assume that the ex-ante choices of the initial labour intensity of production on the newest vintage can be described using a linearly homogeneous production function, which incorporates a term reflecting the influence of embodied technological progress. We furthermore assume that technological progress is (Hicks) neutral and that the production function is separable with respect to labour and capital, on the one hand, and technological progress and effective working hours, on the other. Denoting the time of installation of a certain vintage by T, the ex-ante production function can be described as follows:

$$Q_{T,T} = A(1+g)^T h_T . F(K_{T,T}, L_{T,T}) \tag{3.2.1}$$

where $Q_{T,T}$ is the productive capacity of vintage $T$ at time T , $L_{T,T}$ is the capacity labour demand connected with vintage $T$ at T and $K_{T,T}$ the amount of capital belonging to vintage $T$ at T. $h_T$ is the index of working hours at time T, $g$ is the rate of embodied technical progress[9] and $A$ a positive scalar.

Since the ex-ante production function is assumed to be linearly homogeneous, it follows that equation (3.2.1) can be rewritten in terms of the initial labour coefficient $l_{T,T}$ of a vintage and its initial capital coefficient $k_{T,T}$ as follows:

$$1 = A(1+g)^T h_T . F(k_{T,T}, l_{T,T}) = f(T, l_{T,T}, k_{T,T}) \tag{3.2.2}$$

where $k_{T,T} = K_{T,T}/Q_{T,T}$

and $l_{T,T} = L_{T,T}/Q_{T,T}$

Equation (3.2.2) describes the efficient combination of (initial) labour and capital coefficients as implied by the ex-ante production function. Since we have assumed that entrepreneurs are not myopic with respect to the profitability of a new vintage, i.e. they are interested in the associated returns over a longer period of time, we will now turn to a brief description of the ex-post structure of the vintage model. In this ex-post part we describe what happens to a vintage once its initial labour and capital coefficients have been determined. But since what happens after installation with the returns on a vintage are important from the point of view of maximising total surplus, the ex-post structure determines the ex-ante decision for a large part. Also, while labour and capital productivity are initially determined not only by the choice of a certain labour/capital combination but also by embodied technical progress, this does not imply that labour and capital coefficients remain fixed once their initial values have been determined. Indeed, changes in working hours and in machine time, as well as disembodied technical progress have an influence of their own.[10] Assuming the rates of labour and capital augmenting disembodied technical progress to be equal to $g_L$ and $g_K$[11] respectively, the ex-post development over time of labour and capital requirements per unit of capacity output can be written.

$$l_{T,t} = l_{T,T} \,.\, (h_T/h_t)(1 + g_L)^{-(t-T)} \tag{3.2.3}$$

$$k_{T,t} = k_{T,T} \,.\, (h_T/h_t)(1 + g_K)^{-(t-T)} \tag{3.2.4}$$

$$\mathrm{K}_{T,t} = (1 - d)^{t-T}\, \mathrm{K}_{T,T} \tag{3.2.5}$$

with equation (3.2.5) describing the technical ('radioactive') decay of physical capital.

A quite usual assumption with regard to ex-ante behaviour of producers is that they try to maximise the (expected) net present value of the rent of the newest vintage to be earned over a certain period of time, say z years.[12]. The expected net present value of the rent of vintage $T$, $PV_T^e$ for short, can be written as follows:

$$PV_T^e = \sum_{i=T}^{T+z-1} (p_i^e - w_i^e l_{T,i}^e)\, \mathrm{Q}_{T,i}^e \left(\frac{1}{1+r(T)}\right)^{i-T} - p_K(T)K_{T,T} \tag{3.2.6}$$

where $w^e_i$ is the expected wage rate for year i, r(T) is the ruling discount rate at time T and $p_K(T)$ the price index of new capital at time T.

Note that the net present value of the rent over a period of z years is evaluated at full capacity operation of the vintage in question. Furthermore, if expectations are fully taken into account, the expected period during which a vintage can be profitably operated is effectively a function of the initial labour coefficient of the vintage in question. We will assume z to be exogenously determined for reasons of simplicity. We are at this stage not interested in the absolute value of $PV_T^e$ *per se*, but in its value per unit of initial capacity output which, for short, we will call $pv_T^e$. Using equations (3.2.4) and (3.2.5) we may derive:

$$\mathrm{Q}_{T,t}^e = \mathrm{Q}_{T,T}(h_t^e/h_T)[(1 + g_K)(1 - d)]^{t-T} \tag{3.2.7}$$

Substitution of equations (3.1.2) and (3.2.7) into equation (3.2.6) yields:

$$\begin{aligned} pv_T^e = \frac{PV_T^e}{Q_{T,T}} &= p_T \sum_{i=T}^{T+z-1} a_{T,i} - w_T l_{T,T} \sum_{i=T}^{T+z-1} b_{T,i} - p_K(T)k_{T,T} \\ &= p_T\alpha_{T,z} - w_T l_{T,T}\beta_{T,z} - p_K(T)k_{T,T} \end{aligned} \tag{3.2.8}$$

where

$$a_{T,i} = \left[\frac{(1+\dot{p}_T^e)(1+\dot{h}_T^e)(1+g_K)(1-d)}{1+r(T)}\right]^{i-T}, \quad b_{T,i} = a_{T,i}\left[\frac{1+\dot{w}_T^e}{(1+g_L)(1+\dot{p}_T^e)(1+\dot{h}_T^e)}\right]^{i-T}$$

and

$$\alpha_{T,z} = \sum_{i=T}^{T+z-1} a_{T,i}, \quad \beta_{T,z} = \sum_{i=T}^{T+z-1} b_{T,i}$$

Maximisation of (3.2.8) by choosing $l_{T,T}$ and $k_{T,T}$ subject to equation (3.2.2) yields as first order conditions:

$$-w_T\beta_{T,z} + \lambda \frac{\partial f}{\partial l_{T,T}} = 0 \qquad (3.2.9.a)$$

$$-p_K(T) + \lambda \frac{\partial f}{\partial k_{T,T}} = 0 \qquad (3.2.9.b)$$

$$f(T, l_{T,T}, k_{T,T}) = 1 \qquad (3.2.9.c)$$

where $\lambda$ is the Lagrange multiplier.

Because of the linear homogeneity of $f(T, l_{T,T}, k_{T,T})$ we can derive from (3.2.9a),(3.2.9b) and (3.2.9c):

$$\lambda = w_T\beta_{T,z}l_{T,T} + p_K(T)k_{T,T} \qquad (3.2.10)$$

Equation (3.2.10) states that the Lagrange multiplier (the 'shadow price' of production) equals the present value of the expected cost of obtaining and operating one unit of capacity over a period of z years.

Once the initial values of both the labour coefficient and the capital coefficient have been obtained, we need only observations on the volume of gross investment in order to determine capacity output and capacity labour demand at the vintage level, since both capacity output and capacity labour demand can be determined in a recursive way from the investment figures and the chosen labour and capital coefficients:

$$K_{T,t} = (1-d)^{t-T} K_{T,T} \qquad (3.2.11)$$

$$Q_{T,T} = K_{T,t} / k_{T,t} \qquad (3.2.12)$$

$$L_{T,t} = Q_{T,t} \, . \, l_{T,t} \qquad (3.2.13)$$

Furthermore, assuming that the set of vintages which can be operated at time t is equal to $V_t$, we may calculate aggregate capacity output and aggregate capacity labour demand as:

$$Q_t = \sum_{v \varepsilon V_t} Q_{v,t} \qquad (3.2.14)$$

$$L_t = \sum_{v \varepsilon V_t} L_{v,t} \qquad (3.2.15)$$

### *3.3 Economic scrapping*

#### *3.3.1 The quasi-rent condition*

The ex-post quasi-rent per unit of capacity output of a vintage *T* at time t, $R_{T,t}$, can be written as:

$$R_{T,t} = p_t - w_t l_{T,t} \qquad (3.3.1.1)$$

Once a vintage is installed, it generates these quasi-rents during its entire technical lifetime, which, in our model, is of infinite length. But, if what is happening to the real wage rate now is in some way representative for the future, then producers can increase profits by scrapping a vintage which will generate a negative quasi-rent at the new real wage rate and replacing it by the newest type of vintage. This behaviour is however limited, and can be shown to be inconsistent with the profit maximisation assumption if technological progress is, at least in part, of the embodied kind. This is because the profitability of existing vintages and, hence, the decision to operate or not to operate the vintages in question is evaluated without any reference to the other vintages, including the newest vintage. While, therefore, for an individual vintage the rent over its operating lifetime reaches a maximum value when the associated quasi-rent falls to zero, using this as a scrapping rule for older vintages will not guarantee maximum profits over the entire vintage set. In a profit maximisation setting, the decision on whether to operate a particular vintage or not and the associated scrapping rule need to take account of the profits to be gained from transferring production from profitable existing vintages to more profitable new ones.

### *3.3.2 The Malcomson scrapping condition*

As was mentioned in paragraph 2.2 above, we assume that producers not only decide on the optimum composition of a vintage in terms of labour and capital but also, and perhaps more importantly, on the composition of the capital stock in terms of individual vintages. As has been pointed out by Malcomson,[13] albeit in a different setting, profit maximisation may result in scrapping of equipment which could have earned a positive quasi-rent when operated. Such a kind of scrapping behaviour can be illustrated as follows. Suppose that at time *n*, there are *n* vintages numbered from 1 to *n*. Vintage *n* is assumed to be the youngest one, the size of which still has to be decided upon. Suppose also that producers at time *n* expect to need a volume of total capacity output equal to $Q_n$, and that they invest in new equipment and scrap existing equipment in accordance with this capacity restriction, implying:

$$Q_n = \sum_{i=1}^{n-1} s_i Q_{i,n} + Q_{n,n} \tag{3.3.2.1}$$

where $Q_n$ is the total productive capacity at time *n*, $Q_{i,n}$ is the productive capacity of vintage i at time *n* before economic scrapping and after technical depreciation; $1\text{-}s_i$ is the proportion of productive capacity of vintage i which is scrapped on economic grounds at the beginning of period *n*.

Assuming that the rent per unit of capacity output at time t of vintage *i* is equal to $R_{i,t}$, and that the cost of scrapping capacity belonging to vintage *i* per unit of capacity output is equal to $C_{i,t}$,[14] the current rent of the total capital stock at time n, $R_n$, can be written:

$$R_n = \sum_{i-1}^{n-1} [s_i R_{i,n} - (1 - s_i) C_{i,n}] Q_{i,n} + R_{n,n} Q_{n,n} \tag{3.3.2.2}$$

From equation (3.3.2.1) it follows that:

$$Q_{n,n} = Q_n - \sum_{i-1}^{n-1} s_i Q_{i,n} \tag{3.3.2.3}$$

Substitution of equation (3.3.2.3) into (3.3.2.2) and some rearranging gives the following result for the current total rent:

$$R_n = \sum^{n-1} \{s_i[R_{i,n} - (R_{n,n} - C_{i,n})] - C_{i,n}\} Q_{i,n} + R_{n,n} Q_n \tag{3.3.2.4}$$

Since, by assumption, capital is homogeneous at the vintage level, it follows that $s_i$ is either equal to 1, or $s_i$ equals 0, for all $i \geqslant 1$ and $i \leqslant n - 1$.[15] Since, furthermore, the rent per unit of capacity output on the newest vintage is independent of the size of the newest vintage due to the linear homogeneity of the ex-ante production function, it also follows that equipment of vintage *i* will only be scrapped, i.e. $s_i = 0$, *if* $R_{i,n} < R_{n,n} - C_{i,n}$ or, alternatively, $R_{n,n} - R_{i,n} > C_{i,n}$: existing vintages are replaced by new capacity when the rent differential is large enough to cover the associated scrapping costs. This follows first from the fact that at the ruling prices and given the capacity restriction, the variables $s_i$ are the only choice variables influencing the total rent of the stock of capital, and second from the fact that the negative influence on the value of the total current rent $R_n$ in a situation where $R_{n,n} - R_{i,n} > C_{i,n}$ can be avoided by setting $s_i$ equal to zero. Hence, the set of vintages which are operated can be written:

$$V_t = \{V_{t-1} \cup \{t\}\} \cap \{i \mid i \leqslant t \text{ and } R_{t,t} - R_{i,t} \leqslant C_{i,t}\} \qquad (3.3.2.5)$$

The set intersection present in equation (3.3.2.5) reflects the assumption that vintages which are scrapped will be 'lost' forever. From equation (3.3.2.5) it follows that embodied technological progress has its consequences for the vintage composition of the capital stock. Embodied technological progress alters, *ceteris paribus*, the rent to be earned on the newest vintage, and has therefore a direct influence on the rent differentials. This by itself may be enough to replace otherwise profitable equipment by new (and more profitable) equipment.

### 3.3.3. *Replacement costs*

In order to derive scrapping costs or replacement costs, we concentrate on the change in profitability of producing a certain amount of output using different technologies, i.e. existing ones and the newest one. To this end we have to make some simplifying assumptions first about the way in which investment projects are financed and about capital consumption allowances, and second about the 'value' of machinery and equipment which is scrapped. We assume that:

A. All investment is debt financed.
B. Debt is incurred for a fixed number of years.
C. This number of years is equal to the time period during which capital consumption allowances may be made.
D. Each year a fixed proportion of the initially outstanding debt is repaid. This proportion is equal to the inverse of the number of years for which debt is incurred.
E. Capital consumption allowances follow a linear scheme and they are based on the historic cost-price of investment (corrected for investment grants).
F. Interest payments on debt are tax deductable.
G. The financial consequences of an investment are independent of the actual operation of the machinery in question.
H. The value of machinery and equipment which is laid off due to economic scrapping is equal to zero.

I. Debt is incurred at the beginning of a period, while interest payments, capital consumption allowances and debt repayments are made at the end of the period, with one period intervals.

We will now reformulate the model in order to introduce the corporate tax rate ø as well as the supposed financing of investment in the vintage framework.

Let $D_{T,t}$ be the capital consumption allowances associated with vintage T at time t. Furthermore, let $\gamma$ be the proportion of the price of capital goods which is covered by investment grants.[16] In addition, $t'$ is used to indicate the length of time during which capital consumption allowances, etc., are made. Let $B_{T,t}$ be the outstanding debt associated with vintage T at the end of time period t. Then the after tax present value at time T of the cost of capital at time t associated with vintage T, $PVC_{T,t}$ can be written as:

$$PVC_{T,t} = \left[ r(T)B_{T,t}(1-ø) + \frac{B_{T,T}}{t'} - øD_{T,t} \right] \left( \frac{1}{1+r(T)} \right)^{t-T+1} \quad \text{for } T \leqslant t < T+t' \qquad (3.3.3.1)$$

$$= 0 \quad \text{for } t \geqslant T+t'$$

Where the term $øD_{T,t}$ is the net increase in after tax profits at time t on account of capital consumption allowances associated with vintage T. From assumptions A and E we have:

$$D_{T,t} = B_{T,T}/t' \quad \text{for } T \leqslant t < T+t' \qquad (3.3.3.2)$$

$$= 0 \quad \text{for } t \geqslant T+t'$$

and from assumptions D and I we derive:

$$B_{T,t} = B_{T,T}\left(1 - \frac{t-T}{t'}\right) \quad \text{for } T \leqslant t < T+t' \qquad (3.3.3.3)$$

$$= 0 \quad \text{for } t \geqslant T+t'$$

Furthermore:

$$B_{T,T} = (1-\gamma)p_K(T)K_{T,T} \qquad (3.3.3.4)$$

and using equations (3.3.3.1) to (3.3.3.4) we may now write:[17]

$$PVC_{T,t} = (1-\gamma)(1-ø)p_K(T)\mathrm{K}_{T,T}\left[ r(T)\left(1 - \frac{t-T}{t'}\right) + \frac{1}{t'} \right]\left(\frac{1}{1+r(T)}\right)^{t-T+1} \qquad (3.3.3.5)$$

The total present value of the after tax cost of capital per (initial) unit of capital during the period T to T+t′−1, $TPVC_{T,T+t'-1}$ can then be written as:

$$TPVC_{T,T+t'-1} = (1-\gamma)(1-ø)p_K(T)\sum_{i=0}^{t'-1}\left[ r(T)\left(1 - \frac{i}{t'}\right) + \frac{1}{t'} \right]\left(\frac{1}{1+r(T)}\right)^{i+1} \qquad (3.3.3.6)$$

$$= p_K(T)(1-\gamma)(1-ø)\delta_{T,t'}$$

where

$$\delta_{T,t'} = \sum_{i=0}^{t'-1}\left[ r(T)\left(1 - \frac{i}{t'}\right) + \frac{1}{t'} \right]\left(\frac{1}{1+r(T)}\right)^{i+1}$$

and where we have substituted $i = t - T$.[18]

Taking account of the influence of the corporate tax rate on the after tax net present value of an investment project we can rewrite (3.2.8) as:

$$pv_T^{\text{¢}} = (p_T\alpha_{T,z} - w_T l_{T,T}\beta_{T,z})(1-ø) - (1-\gamma)(1-ø)p_K(T)k_{T,T}\delta_{T,t'} \qquad (3.3.3.7)$$

The first-order conditions of the revised maximisation problem now result in:

$$\frac{\partial f/\partial l_{T,T}}{\partial f/\partial k_{T,T}} = \frac{w_T\beta_{T,z}}{(1-\gamma)p_K(T)\delta_{T,t}} \qquad (3.3.3.8)$$

Equation (3.3.3.8) together with equation (3.2.9c) provides us with the initial values of the labour and capital coefficients, which solve the net present value maximisation problem for a fixed planning period z and a fixed capital consumption allowance period t′. Since $\delta_{T,t'} = 1$ and as the influence of the corporate tax rate vanishes, the solution of the revised maximisation problem is different from the solution of the unrevised problem only if $\gamma \neq 0$.

Having specified the nature of the capital costs associated with installing a vintage, we will now turn to the introduction of these costs in the scrapping condition. It is clear that the returns on a vintage which has been installed at time $T$ and which is operated at time $t$, where $t > T + t'$, are equal to the quasi-rent generated by operating that vintage. Since we have assumed that the act of scrapping is costless, there is a net profit to be gained from replacing such a vintage by the newest type of equipment if the rent on the newest vintage is larger than the quasi-rent on vintage $T$. This is, however, not that obvious in the case where $t < T + t'$, for, whether vintage $T$ is operated or not, there are still capital costs to be covered from operating the vintage itself, or, perhaps, by operating that part of the newest vintage which replaces vintage $T$.

In order to resolve this problem, it is necessary to be more specific about the meaning of scrapping costs. In this model, scrapping costs or replacement costs are defined as that part of the net profits per unit of capacity output generated by new investment which is absorbed by meeting the financial obligations associated with the old investment projects. Since these capital costs may have to be covered for a longer period of time, it stands to reason that we evaluate the present value of the rent differentials during that period. First, however, we may note that the rent differential condition for a particular year is equivalent to the Malcomson scrapping condition, which states that old machinery and equipment is replaced when the quasi-rent per unit of capacity output to be earned on the old vintage is smaller than the rent per unit of output on the new vintage.[19] This can be illustrated as follows. From equations (3.2.5) and (3.3.3.1) we have:

$$R_{T,t} = \overline{R}_{T,t}(1-ø) - \frac{PVC_{T,t}k_{T,t}}{(1-d)^{t-T}} \qquad (3.3.3.9)$$

where $R_{T,t}$ now equals the after tax rent per unit of output of vintage $T$ at time t. The after tax rent differential condition for time t and for the new vintage $T$ can therefore now be written as:

$$R_{t,t} - R_{T,t} = (1-ø)(\overline{R}_{t,t} - \overline{R}_{T,t}) - \left(PVC_{t,t}k_{t,t} - \frac{PVC_{T,t}k_{T,t}}{(1-d)^{t-T}}\right) > \frac{PVC_{T,t}k_{T,t}}{(1-d)^{t-T}} \qquad (3.3.3.10)$$

where vintage $T$ is scrapped if (3.3.3.10) is true.

Substitution of equation (3.3.1.1) into equation (3.3.3.10) results in the following equivalent representation of the scrapping condition:

$$\frac{w_t(l_{T,t} - l_{t,t})}{p_t} > \frac{PVC_{t,t}k_{t,t}}{(1 - ø)p_t} \tag{3.3.3.11}$$

If we assume that the share of labour on a particular vintage does not decrease over time, and if we further assume that the impact of changes in effective working hours and of disembodied technical progress is not vintage specific,[20] then we may conclude that the quasi-rent differential does not decrease over time. Since, furthermore, the nominal user cost of capital decreases over time due to smaller interest charges,[21] it follows that for time $t' > t$ the quasi-rent differential condition is fulfilled as well. Therefore, in normal cases with a non negative labour share growth on old vintages, it suffices to check for the fulfilment of the quasi-rent differential condition for the current period only, and we may write the set of vintages to be operated as:

$$V_t = \{V_{t-1} \cup \{t\}\} \cap \left\{ i \mid i \leq t \ \wedge \ w_t(l_{i,t} - l_{t,t}) \leq \frac{PVC_{t,t}k_{t,t}}{1 - ø} \right\} \tag{3.3.3.12}$$

## 3.4 *Closing the model*

### 3.4.1 *The ex-ante production function*

Having specified the general outlines of the production model, it is now time to make it operational by selecting a specification for the ex-ante production function. One specification which has been widely used is the Cobb-Douglas production function, or CD function for short. Because this function is given in log linear form, there is no use in separating embodied technological progress into labour and/or capital augmenting technological progress. Indeed, taking exponential functions to represent these different types of technological progress, the CD function simply 'lumps' them together. We therefore cannot tell whether technological progress is mainly labour augmenting, capital augmenting or neutral. While this may be thought of as a drawback connected with using the CD function, it may also be regarded as an advantage in the circumstances we are in, i.e. suffering from lack of observations. For reasons of simplicity, we therefore assume that technological progress is neutral.

Another drawback, which may be more serious in nature, is that in a CD function the implied elasticity of substitution between labour and capital is equal to 1. This means that a 1 per cent change in the ratio of labour cost to capital cost per unit of labour and capital, respectively generates a 1 per cent (compensating) change in the optimum capital/labour intensity of production. A CD function may therefore exaggerate the possibility of factor substitution at the aggregate level. By breaking down aggregate capacity into 'putty-clay' vintages, however, the direct substitution effect of a change in the relative price of labour and capital is noticed on the newest vintage only. Whereas, therefore, the substitution effect as implied by an aggregate CD production function is relatively important, the same functional specification of the production function in a putty-clay vintage setting generates less-pronounced substitution effects, due to the fact that substitution in the broad sense of the word, i.e. including the substitution of labour intensive vintages by more capital intensive ones, does not affect the entire stock of capital.[22]

The choice of a CD function makes it fairly easy to complete the model as it stands. The initial labour and capital coefficients can be found from equation (3.3.3.8) and by substitution of the CD function into equation (3.2.2):

$$k_{T,T} = [A\,(1+g)^T h_T l_{T,T}^{\mathrm{d}}]^{\frac{-1}{1-\alpha}} \tag{3.4.1.1}$$

where α is the partial output elasticity of labour.

From equation (3.3.3.8) we find:

$$l_{T,T} = k_{T,T}\,\frac{\alpha}{1-\alpha}\,\frac{(1-\gamma)p_K(T)}{w_T\beta_{T,z}} \tag{3.4.1.2}$$

Combining equations (3.4.1.1) and (3.4.1.2) we arrive at:

$$k^*_{T,T} = \left(\frac{1-\alpha}{\alpha}\,\frac{w_T\beta_{T,z}}{(1-\gamma)p_K(T)}\right)^{\alpha} \frac{1}{A(1+g)^T h_T} \tag{3.4.1.3}$$

$$l^*_{T,T} = \left(\frac{1-\alpha}{\alpha}\,\frac{w_T\beta_{T,z}}{(1-\gamma)p_K(T)}\right)^{\alpha-1} \frac{1}{A(1+g)^T h_T} \tag{3.4.1.4}$$

Equations (3.4.1.3) and (3.4.1.4) relate the optimum technical coefficients to variables which can, in principle at least, be observed.

### *3.4.2 Capacity labour demand and employment*

Presumably time series on capacity labour demand are not available. We therefore have to proceed linking this unobserved variable to an observable variable, and employment is the obvious candidate. We first introduce the rate of labour utilisation, $q^n_t$, which is equal to:

$$q^n_t = E_t/L_t \tag{3.4.2.1}$$

Where $E_t$ equals total employment at time t and $L_t$ is equal to total capacity labour demand at time *t*. From equation (3.4.2.1) it follows that:

$$E_t = q^n_t . L_t \tag{3.4.2.2}$$

Presumably, older vintages are more labour intensive than younger vintages.[23] Therefore, the minimum required rate of labour utilisation, $q^{n*}_t$, which is defined as the rate of labour utilisation which is needed to produce actual output, is less than or equal to the actual rate of capacity utilisation, $q_t$.[24] Therefore, if the rate of capacity utilisation increases, more labour intensive machinery and equipment will have to be utilised and a change in the demand for output will generate a more than proportional change in required employment, $E^*_t$, which can now be written:

$$E^*_t = q^{n*}_t . L_t \tag{3.4.2.3}$$

Actual employment will be at least equal to required employment. It may be larger than required employment if the costs of hiring and firing are positive. We may model the latter by assuming that actual employment is equal to the sum of ex post required employment and a fraction ζ of the difference between ex-ante desired and ex-post required employment, giving:

$$E_t = E_t^* + \zeta \cdot (L_t - E_t^*) \tag{3.4.2.4}$$

Where $0 \leqslant \zeta \leqslant 1$. From equations (3.4.2.3) and (3.4.2.4) it follows that:

$$E_t = (q_t^{n*} + \zeta \cdot \{1 - q_t^{n*}\}) \cdot L_t \tag{3.4.2.5}$$

Equation (3.4.2.5) relates changes in capacity labour demand as generated by the vintage model to observable changes in employment, using a simple 'error' adjustment scheme. The cost of not having capacity labour demand time series available is clear. We need to extend the model with an employment function expanding the parameter set by at least one parameter, i.e. $\zeta$.

## *3.5 Technological progres and R&D*

### *3.5.1 Introduction*

In this section we will give some attention to the way in which R&D has been used to bridge the gap between the growth of output and that of factor productivity. As was mentioned in Section 2, the influence of R&D on the growth of output has been looked at in a manner which is not that different from the way in which labour and capital are considered to influence output growth. Indeed, Mansfield and Griliches,[25] for instance, use the concept of an R&D capital stock, and variations in this stock, to provide an explanation for the technological shift of an otherwise 'normal' CD production function. In the next subsection (3.5.2) we will look into this matter in somewhat more detail, while in 3.5.3, we will try to introduce our view with respect to impact of R&D on productivity growth into the basic structure of the putty-clay vintage model as outlined in the previous sections.

### *3.5.2 R&D stocks and flows*

As Griliches points out, 'Past research and development investments depreciate and become obsolete. Thus the growth in the net stock of research and development capital is not equal to the gross level of current or recent resources invested in expanding it.'[26] This seems to be a statement that is easy to accept. But the acceptance or (partial) rejection depends on the specific framework used in order to incorporate the notion of an R&D stock. Indeed, depreciation and obsolescence seem to be connected with 'earning capacity' rather than with something which actually happens to the physical stock of R&D. Looked at in this way, the influence of R&D on productivity is related to money productivity changes rather than 'real' productivity changes, i.e. R&D investors cannot reap the monetary benefits of their R&D investments for more than a limited number of years.

The question to be addressed is whether or not this observation is relevant within a production function setting. For, in such a setting it is potential (and real) output which is related to the most efficient combinations of real labour input, real capital input and (productive) R&D stock services. Within the aggregate production function framework there is reason to question the validity of Griliches's assertion that 'We can say then, equivalently, that either only a fraction of current research and development flow is to be thought of as a net addition to the social stock of knowledge capital or that some faction of the pre-existing stock of this capital is replaced (depreciated) annually. The real problem here is our lack of information about the possible rates of such depreciation.'[27] In our opinion the latter is not the real problem. The real problem is the supposed equivalence between (an automatic?) depreciation (scheme) and a net addition to the R&D stock which is smaller than current/recent R&D efforts. That is not to say that R&D obsolescence does not occur, for this is precisely the reason why R&D is undertaken: finding ways to do better or products which are better. Taking the equivalence assumption to the extreme as do Patel and Soete,[28] it can be argued that the 'approximate' equivalence breaks down for high rates of depreciation and low levels of current R&D, in which case the R&D stock of capital, and therefore its productive content, may even decrease. For it seems hard to imagine that R&D results are implemented irrespective of their performance relative to the performance of the R&D findings they are meant to replace. If the new results are better than the old ones, then the implementation of the new results should lead to a net increase of the productive content of the R&D stock of capital. If not, they are just not implemented and the R&D stock does not change at all, and potential output is either positively affected or not at all.[29]

The question which can now be raised is whether the usual stock concept is a useful one. While we have put forward a number of objections against using this stock concept and its associated depreciation scheme, there does not really seem to be an alternative which is acceptable from a practical point of view. For the replacement or obsolescence of parts of the R&D stock cannot be handled in a proper manner without being more specific about quality differences in current and past R&D, for instance. Because we do not have any information with regard to these differences, we cannot say anything about the rate of obsolescence of R&D based on differences in the productive contents of different parts of the stock of R&D, except perhaps that it is non-negative. In addition, the successful implementation of 'net' 'own' R&D efforts may very well depend on the quality of the stock of capital (and therefore on embodied R&D), in which case the 'equivalent' rate of obsolescence of the stock of 'own' R&D is influenced by the composition of the stock of physical capital in terms of individual vintages: the stock of 'own' R&D capital may therefore not be an independent variable.[30] While, therefore, we basically disagree with the 'common practice' of adding recent and current real R&D expenditures together in order to arrive at a measure of the 'stock' of R&D as if real R&D efforts are homogeneous in time,[31] we have to proceed along similar lines. We will however assume an R&D stock construction scheme which implicitly allows for a zero rate of depreciation of the stock of R&D. We will present this construction scheme in the next subsection.

### *3.5.3 R&D stock construction*

As Griliches has pointed out, there are a number of lags involved between R&D expenditures and the actual availability or use of the accompanying results.[32] Furthermore, the 'bell shaped curve' seems to be a realistic a priori notion about the distribution of the lags involved for the transformation R&D expenditures into additions to the stock of R&D capital. For reasons of both simplicity and parameter parsimoneousness, we opt for a binomial lag distribution, giving:[33]

$$R\,\&\,D_t = \sum_{i=0}^{n} \left( \frac{n\,!}{(n-i)\,!\,i\,!} \mu^i (1-\mu)^{n-i} \right) r\,\&\,d_{t-i} \tag{3.5.3.1}$$

where $R\,\&\,D_t$ equals the addition to the stock of R&D at time t, $r\,\&\,d_t$, equals real R&D expenditures at time t, n is the longest lag with a positive weight in the binomial distribution, and μ is the binomial distribution parameter.

If μ is equal to 0.5, then the lag distribution is symmetric. If $\mu < 0.5$ then recent R&D expenditures have a relatively large impact on the R&D stock: the lag distribution is skewed towards more recent R&D.

Superimposed on the 'bell shaped' binomial lag distribution, is another distribution describing the volume of the current stock of R&D capital in terms of its composing elements, i.e. additions to the stock both in the present and in the past. Let us assume that these weights are exponentially declining, as is the case in perpetual inventory constructs. In this case the stock of R&D capital at time t, $M_t$ can be written:

$$M_t = R\,\&\,D_t + (1-\delta) \cdot M_{t-1} = \sum_{i=n+1}^{t} R\,\&\,D_i \cdot (1-\delta)^{t-i} + (1-\delta)^{t-n} \cdot M_n \tag{3.5.3.2}$$

where δ is equal to the rate of depreciation of the stock of R&D. Denoting the weights of the binomial lag function by $W_i$ and the exponentially declining weight for time t by $U_t$, the stock of R&D capital at time t can be written:

$$M_t = \sum_{i=n+1}^{t} U_i \sum_{j=0}^{n} r\,\&\,d_{i-j} \cdot W_j + (1-\delta)^{t-n} \cdot M_n \tag{3.5.3.3}$$

The effective weights of real R&D expenditures in the determination of the stock of R&D capital are therefore made up of a number of products of the exponential weights and the binomial weights.[34] From equation (3.5.3.3) it follows that for large enough *t–n* and a positive rate of depreciation δ, the influence of $M_n$ on $M_t$ may be disregarded. Hence, the stock of R&D capital can be written as a weighted sum of past Rr&D expenditures where the weights are again approximately distributed following a bell shaped pattern. If n is smaller than the half life of the R&D stock, and this seems to be a reasonable assumption within this particular framework, then the distribution pattern will be dominated by the exponentially declining weights, giving rise to a bell shaped distribution which is skewed towards more recent R&D expenditure. If, on the other hand, the distribution proves to be symmetric, we may conclude that apparently depreciation of the R&D stock is of negligible importance, and that the overall distribution is approximately uniform.[35] As an approximation to the effective weights of past and current R&D expenditures in the current stock of R&D capital we may use a binomial distribution function, which incorporates a skewed bell shape, as was mentioned above. If the rate of depreciation proves to be negligible, i.e. the distribution of the weights is more or less symmetric, we may not neglect the influence of the initial stock of capital, in which case we need an extra parameter, i.e. $M_n$, in order to construct the stock of R&D capital using equation (3.5.3.2) with $\delta=0$.

We assume now that the rate of growth of R&D services is proportional to the rate of growth of the stock of R&D capital, and that furthermore the 'rate of technological progress' $g_t$ at time t is equal to the rate of change of R&D services and a constant trend term $g'$ reflecting systematic changes in productivity not captured by the R&D stock construct, i.e.:

$$1+g_t=\left(1+f^e\cdot\frac{M_t-M_{t-1}}{M_{t-1}}\right)\cdot(1+g') \tag{3.5.3.4}$$

Where $f^e$ is a parameter comparable with the partial output elasticity of the stock of R&D capital in a Cobb-Douglas production function. Using equation (3.5.3.4), we can now define a technological shift function $S_t$ as follows:

$$\begin{aligned} S_t &= [1+g_t]S_{t-1} && t>0 \\ &= 1 && t=0 \\ &= S_{t+1}/(1+g_{t+1}) && t<0 \end{aligned} \tag{3.5.3.5}$$

Using equation (3.5.3.5) and a priori values or 'estimates' of $\mu$ and $g'$, we can build a technological shift time series which can easily be introduced both into the CD ex-ante function of the putty clay model and the Leontief ex-post functions. We will further elaborate on this in the next subsection.

### *3.5.4 R&D productivity and vintage capital*

Having specified the technological shift function, we now turn to the incorporation of this function in the vintage capital framework. In order to do so, we have to define 'embodied' and 'disembodied' technological progress in terms of the R&D based shift functions. In the following, embodied technological progress is associated with the rate of growth of the productive services of the stock of R&D capital in the investment goods industry, while 'disembodied' technological progress is associated with the rate of growth of productive services rendered by the stock of 'own' R&D. Both rates of technological progress contain a term representing systematic productivity changes over time not captured by the R&D stock based productivity index constructs.

We first change the ex-ante CD function in terms of these new technological progress functions. In effect we assume that the total technological shift consists of the product of two shift functions, the first being associated with 'embodied technological progress', $S_T^e$, and the second one with 'disembodied' technological progress, $S_T^s$.[36] [37] Both types of technological progress are assumed to be neutral with respect to their impact on the initial labour and capital coefficients. Hence, the ex-ante function changes into:

$$Q_{T,T}=A\, S_T^e S_T^s h_T\cdot F(\mathrm{L}_{T,T},\mathrm{K}_{T,T}) \tag{3.5.4.1}$$

After installation of the vintage, only disembodied technological progress has an effect on vintage productivity, giving:

$$l_{T,t}=\frac{l_{T,T}(h_T/h_t)}{(S_t^s/S_T^s)\cdot(1+g'_L)^{t-T}} \qquad \text{for} \quad t\geqslant T \tag{3.5.4.2}$$

$$k_{T,t} = \frac{k_{T,T}(h_T/h_t)}{(S_t^s/S_T^s)\cdot(1+g'_K)^{t-T}} \qquad \text{for} \quad t \geqslant T \tag{3.5.4.3}$$

Note that these specifications resemble the ex-post technological coefficients specifications as described in equations (3.2.3) and (3.2.4) very closely. Furthermore, in the case of the Netherlands for instance, there is evidence that disembodied technical progress is mainly labour augmenting.[38] We may test for the presence of such a bias by excluding the technological shift function from equation (3.5.4.3) as a separate version of the model.[39]

In order to bring them more in line with casual observation, however, it may be necessary to alter equations (3.5.4.2) and (3.5.4.3) somewhat. For, whereas disembodied technological progress falls as 'manna from heaven' and is not vintage specific, process and product R&D as performed within the sector of industry may not have this 'quality of indifference'. When we think of the ex-ante isoquant as consisting of a series of different labour and capital combinations which are best suited to implement 'own' process and product R&D efforts, then it stands to reason that new products and processes can be implemented on existing machinery only when the 'technological distance' between the process/product 'embodied' and the process/product to be implemented is fairly small.[40] We might even try to implement this casual observation into the model by postulating:

$$g_{T,t}^s = f^s \cdot \left[\frac{M_t^s - M_{t-1}^s}{M_{t-1}^s}\right] \cdot \left[\frac{S_T^e}{S_t^e}\right]^{\Gamma} \tag{3.5.4.4}$$

where $g^s_{T,t}$ is the rate of growth of technological progress due to process and product R&D within a particular sector of industry for vintage T at time t, and $M^s_t$ is the stock of 'own' R&D within a particular sector of industry. Note that $\Gamma=0$ implies that the impact of disembodied technological progress on labour and capital productivity is not vintage specific. If $\Gamma>0$ then embodied and disembodied technological progress have a more or less complementary character. Incorporating this complementary notion into equations (3.5.4.2) and (3.5.4.3) by using equation (3.5.4.4), we arrive at:

$$l_{T,t} = \frac{l_{T,t-1}h_{t-1}/h_t}{(1+g_{T,t}^s)(1+g'_L)} \tag{3.5.4.5}$$

$$k_{T,t} = \frac{k_{T,t-1}h_{t-1}/h_t}{(1+g_{T,t}^s)(1+g'_K)} \tag{3.5.4.6}$$

Equations (3.5.4.5) and (3.5.4.6) describe the time paths of the labour coefficient and the capital coefficient of vintage T as they are influenced by changes in effective working hours and changes in the rates of growth of labour and capital productivity induced by 'own' R&D. Because of this influence, the time path does not only depend on expectations with respect to working hours but also on plans and realisations of 'own' R&D. Therefore, we have to make a distinction between the ex-ante time path of labour productivity and capital productivity as influenced by R&D plans, and the ex-post time path of labour and capital productivity as the result of the realisation of these plans. For reasons of simplicity, we may regard actual R&D efforts as identical to R&D plans, in order to be able to calculate the net present value of an investment project, where possible. For vintages which are installed at a time near the end of the estimation period, we may need observations on future R&D expenditures which are not available. In this case we will assume that plans are such that a constant rate of 'own' R&D based technological progress will be generated.[41]

### 3.6 *The vintage model: a brief summary and some concluding remarks*

#### 3.6.1 *Summary of the model*

In this section we will briefly summarise the equations of the model as it has been described in the previous paragraphs. In addition, we define an equation for the set of vintages which are exactly sufficient (ex post) to produce actual output in the most profitable way, i.e. using the least possible amount of labour. This particular set functions as an argument in the employment equation. The model consists of the following equations:

$$M_t^{s,e} = \sum_{i=0}^{n^{s,e}} \left( \frac{n^{s,e}\,!}{(n^{s,e}-i)\,!\,i\,!} \mu_{s,e}^{i} \, (1-\mu_{s,e})^{n^{s,e}-i} \right) r\,\&\,d_{t-i}^{s,e} \quad \text{for } \mu < 0.5 \tag{3.6.1.1}$$

$$R\,\&\,D_t^{s,e} = \sum_{i=0}^{m^{s,e}} \left( \frac{m^{s,e}\,!}{(m^{s,e}-i)\,!\,i\,!} \nu_{s,e}^{i} \, (1-\nu_{s,e})^{m^{s,e}-i} \right) r\,\&\,d_{t-i}^{s,e} \quad \text{for } \mu \sim 0.5 \tag{3.6.1.2}$$

$$M_t^{s,e} = M_{t-1}^{s,e} + R\,\&\,D_t^{s,e} \quad \text{for } \mu \sim 0.5 \tag{3.6.1.3}$$

$$1 + g_t^e = \left( 1 + f^e \cdot \left\{ \frac{M_t^e - M_{t-1}^e}{M_{t-1}^e} \right\} \right) \cdot (1 + g^{e'}) \tag{3.6.1.4}$$

$$g_{T,t}^s = f^s \cdot \left\{ \frac{M_t^s - M_{t-1}^s}{M_{t-1}^s} \right\} \cdot \left[ \frac{S_T^e}{S_t^e} \right]^{r} \tag{3.6.1.5}$$

$$\begin{aligned} S_t^e &= [1 + g_t^e] S_{t-1}^e && t > 0 \\ &= 1 && t = 0 \\ &= S_{t+1}^e / (1 + g_{t+1}^e) && t < 0 \end{aligned} \tag{3.6.1.6}$$

$$\begin{aligned} S_t^s &= [1 + g_{t,t}^s]\, S_{t-1}^e && t > 0 \\ &= 1 && t = 0 \\ &= S_{t+1}^s / (1 + g_{t+1,t+1}^s) && t < 0 \end{aligned} \tag{3.6.1.7}$$

$$Q_{T,T} = A \cdot S_T^e \cdot S_T^s \cdot h_T \cdot F(L_{T,T}, K_{T,T}) \tag{3.6.1.8}$$

$$\beta_{T,z} = \sum_{i=T}^{T+z-1} \left( \frac{(1 + w_T^e)(1 - d)(1 + g'_K)}{(1 + g'_L)(1 + r(T))} \right)^{i-T} \tag{3.6.1.9}$$

$$l_{T,T} = k_{T,T} \frac{\alpha}{1-\alpha} \frac{(1-\gamma) p_K(T)}{w_T \beta_{T,z}} \tag{3.6.1.10}$$

$$k_{T,T}^{*} = \left( \frac{1-\alpha}{\alpha} \frac{w_T \beta_{T,z}}{(1-\gamma) p_K(T)} \right)^{\alpha} \frac{1}{A S_T^e S_T^s h_T} \tag{3.6.1.11}$$

$$l^*_{T,T} = \left(\frac{1-\alpha}{\alpha} \frac{w_T \beta_{T,z}}{(1-\gamma) p_K(T)}\right)^{\alpha - 1} \frac{1}{A S^e_T S^s_T h_T} \tag{3.6.1.12}$$

$$l_{T,t} = \frac{l_{T,t-1} h_{t-1}/h_t}{(1+g^s_{T,t})(1+g'_L)} \tag{3.6.1.13}$$

$$k_{T,t} = \frac{k_{T,t-1} h_{t-1}/h_t}{(1+g^s_{T,t})(1+g'_K)} \tag{3.6.1.14}$$

$$k_{T,t} = k_{T,T}(1-d)^{t-T} \tag{3.6.1.15}$$

$$Q_{T,t} = K_{T,t}/k_{T,t} \tag{3.6.1.16}$$

$$L_{T,t} = Q_{T,t} \cdot l_{T,t} \tag{3.6.1.17}$$

$$V_t = \{V_{t-1} \cup \{t\}\} \cap \{i \mid i \leqslant t \tag{3.6.1.18}$$

$$\text{and} \quad w_t(l_{i,t} - l_{t,t}) \leqslant (1-\gamma) \cdot p_K(T) \cdot k_{t,t} \cdot \{\frac{r(t) + 1/t'}{1 + r(t)}\}\}$$

$$Q_t = \sum_{v \in V_t} Q_{v,t} \tag{3.6.1.19}$$

$$L_t = \sum_{v \in V_t} L_{v,t} \tag{3.6.1.20}$$

$$V^+_t = \{i\} \mid \{i\} \subseteq V_t \wedge \sum_{k \in \{i\}} Q_{kt} \geqslant X_t \tag{3.6.1.21}$$

where $X_t$ is equal to actual output at t.

$$V^*_t = \left\{\{i\} \mid \{i\} \subseteq V^+_t \wedge \left(\sum_{k \in \{i\}} w_t \cdot l_{k,t} \leqslant \sum_{j \in \{x\}} w_t \cdot l_{j,t} \, \forall \{x\} \subseteq V^+_t\right)\right\} \tag{3.6.1.22}$$

$$E^*_t = \sum_{v \in V^*_t} L_{v,t} \tag{3.6.1.23}$$

$$E_t = E^*_t + \zeta \cdot \{L_t - E^*_t\} \tag{3.6.1.24}$$

### *3.6.2 Concluding remarks*

Having described the impact of R&D based technological progress on both the ex-ante production function and the ex-post development of the labour and capital coefficient, we may observe that changes in R&D efforts affect the stock of capital in two ways. First, an increase in 'embodied' technological progress tends, *ceteris paribus*, to increase the rent differential of the newest vintage and the existing vintages. When therefore R&D efforts within the investment goods industry are increasing, the average age of the capital stock of a particular sector of industry will decrease. Secondly, if $\Gamma > 0$ then 'own' R&D will generate a decrease in the average age of the capital stock as well: in both cases scrapping leads to a decrease of ex-post rent differentials. If these effects prove to be significant, then we may tentatively draw two conclusions with respect to the paradox of the slowdown of the rate of growth of total factor productivity and the increase in recent R&D efforts:

R&D efforts tend to increase the productive contents of the capital stock through an acceleration of economic obsolescence, on the one hand, and

Perpetual inventory constructs of the capital stock tend therefore to overestimate the size of the capital stock, especially in times of relatively large increases in real wages or large scale R&D efforts. In these circumstances, net investment is overestimated (due to the increase in economic obsolescence which is not reflected by the constant depreciation parameter in the perpetual inventory stock construct), and consequently the increase of the capital stock is overestimated, thereby reducing observed changes in total factor productivity.

If R&D efforts have an impact on the vintage composition of the capital stock as described above, then an increase in R&D efforts tends to lower the average age of the capital stock. If the capital stock has become very young due to persistent R&D efforts or persistently high real wage growth, then a situation might arise where R&D induced productivity growth slows down because of the accompanying decrease in rent differentials.

## 4. Summary and concluding remarks

In this chapter we have described a simple vintage model with substitution possibilities between labour and capital ex ante and 'fixed' labour and capital coefficients ex post. Entrepreneurs try to maximise the net present value of the rent associated with installing and subsequently operating a vintage over a fixed number of years. The efficient combinations of labour and capital available to them are described using a linearly homogeneous Cobb-Douglas function. The ex-ante function contains a term reflecting the influence of embodied technological progress. After the installation of a particular vintage, labour and capital coefficients may change due to disembodied technological progress and changes in effective working hours. Because of the linearly homogeneous character of the ex-ante production function, the rent per unit of capacity output on a vintage is independent of the size of the vintage. We use this result in the determination of the optimum composition of the capital stock in terms of individual vintages. Expectations with regard to the future values of the economic variables relevant to the investment decision, are based on current and recent observations on these variables: expectations are adaptive.

Once entrepreneurs have decided on the initial composition of the new vintage in terms of labour and capital requirements per unit of capacity output, they have to decide on the scale of production on the newest vintage. We assume that entrepreneurs are facing a constraint with respect to total capacity output. From the profit maximisation principle, we derive the Malcomson scrapping condition, and it is the scrapping behaviour of entrepreneurs which determines the size of the newest vintage as well as the age of the capital stock in such a way that the resulting profitable capacity is in line with the capacity restriction mentioned above. Furthermore, the capital stock thus determined, generates a maximum value of the (expected) total rent stream.

We proceed with 'linking' the rates of embodied and disembodied technological progress to R&D efforts in the investment goods industry and to 'own' efforts. We construct proxies of the 'stocks' of R&D capital. We also include the notion that the impact of 'own' R&D on the productivity of vintage capital may be vintage specific and thus more or less complementary to the level of embodied technological progress. The formal specification of R&D complementarity is quite simple, but it does allow for complete complementarity, no complementarity at all as well as partial complementarity. The estimation of the parameters of the model would provide a tentative answer to this complementarity question.

The proposed linking of R&D to vintage specific technological progress has some interesting consequences. First of all the age of the capital stock decreases with an increase in 'embodied' and 'own' R&D efforts.[42] Therefore, if we do not allow for the influence of technological progress on the age of the capital stock (for instance, in the context of a perpetual inventory capital stock construct), the rate of increase of the capital stock will be overestimated in times of increasing R&D efforts or in times of sharp increases in real wages. This leads us to the tentative conclusion that the paradox of increasing R&D efforts and a slowdown of total factor productivity growth may be partly due to a 'mis-measurement' of the stock of capital. Furthermore, if the age of the capital stock decreases relatively rapidly, as it should in time of increasing R&D efforts, then the decrease of the age of the capital stock itself will be associated with a decrease in rent differentials, which diminishes the influence of current R&D on the age structure and the productivity of the capital stock. So, while an increase in R&D efforts may lead to an initial spurt of productivity growth, this increase in R&D efforts, even when the latter is a structural phenomenon, loses part of its initial impact on productivity growth because rent differentials will become less pronounced.

## Notes

1. Solow, R. (1987), 'Technical change and the aggregate production function', *Review of Economics and Statistics*, pp. 312–20.
2. For a detailed description of the assumptions and their implications see, for example, Nickell, S. (1978), *The Investment Decisions of Firms*, Oxford.
3. See, for example, Griliches, Z. (1979), 'Issues in assessing the contribution of research and development to productivity growth, *Bell Journal of Economics*, Vol. 10, pp. 92–116, 'Productivity, R&D and basic research at the firm level in the 1970s', *American Economic Review*, 1986, Vol. 76, pp. 141–54; and 'R&D and productivity: measurement issues and econometric results', *Science*, 237, 1987, pp. 31–5. See also Mansfield, E. (1980), 'Basic research and productivity increase in manufacturing', *American Economic Review*, Vol. 70, pp. 863–873.
4. Griliches mentions the distinction between 'partial' contributions of R&D to growth and the total contribution of R&D to growth through (additional) induced changes in the other inputs. But whereas Griliches concentrates on the former type of contribution, we will focus our attention on the latter. Griliches (1979), p. 93.
5. There is another source of heterogeneity, however, which is still ignored in vintage models. Vintages consist of the smallest units of capital by means of which output can be produced: generally a bundle of heterogeneous investment goods, or even entire factories, rather than of some kind of single good. At this micro-structure level vintages are themselves heterogeneous. An additional source of heterogeneity results from the non-instantaneous diffusion of new technologies. Both types of heterogeneity are ignored in this chapter for reasons of simplicity.

6. Other types of vintage models are 'putty-putty' and 'clay-clay'. In the former, capital/labour ratios are fully flexible both before and after a particular vintage has been chosen. In models where capital is of the 'clay-clay' type, different vintages are characterized by different capital/labour ratios which are fixed and predetermined.
7. For a description of these types of models, see for example Wan, H. Jr (1971), *Economic Growth*, New York, chapter 5.
8. See, for instance, Johansen, L. (1959), 'Substitution versus fixed production coefficients in the theory of economic growth: a synthesis', *Econometrica*, 27, pp. 157–75.
9. In a later stage we will redefine the technological shift function to include quality improvements of output, on the one hand, and quality improvements of production processes through 'own' R&D, on the other. For reasons of exposition we will not do the same here, although the rate of technological progress could easily be thought of as consisting of two parts, i.e. a truly embodied part and a disembodied part reflecting the quality improvements mentioned above.
10. Note, however, that later on we will assume that disembodied technical progress consists of two parts: a neutral part (neutral in the sense that its impact on ex-post labour and capital productivity growth is the same); and a specific labour or capital augmenting part. For ease of exposition we will postpone the introduction of this assumption.
11. See note 10.
12. See, for instance, Gelauff, G.M.M., Wennekers, A.R.M. and A.H.M. de Jong, 'A putty-clay model with three factors of production and partly endogenous technical progress', *De Economist*, (1985) Vol. 133, pp. 327–351; Kuipers, S. and A. van Zon, 'Output and employment growth in the Netherlands in the post-war period: a putty-clay approach', *De Economist*, (1982) Vol. 130, pp. 38–70; and Muysken, J. and A. van Zon, 'Employment and unemployment in the Netherlands, 1960–1984: a putty-clay approach', *Recherches Economique de Louvain*, (1987) Vol. 53, pp. 101–33.
13. Malcomson, J.M. (1975), 'Replacement and the rental value of capital equipment subject to obsolescence', *Journal of Economic Theory*, 10, pp. 24–41.
14. In the next subsection we will tentatively define scrapping costs.
15. That is, except for a situation where existing capacity at a certain moment of time is large enough to generate total output and existing capacity can be operated more profitably than new capacity. In these circumstances gross investment will be zero, and the survival fraction of the least profitable vintage may be less then one but larger than zero. When gross investment is not equal to zero, i.e. when new capacity is installed, the survival fractions of the other vintages must either be equal to one or equal to zero. This is caused by the assumption of homogeneity of capital at the vintage level and the logical requirement that the scrapping condition should be internally consistent. When, therefore, a piece of equipment belonging to a certain vintage becomes obsolete, every piece of equipment belonging to that vintage becomes obsolete, and so the entire vintage should be scrapped rather than just part of it.
16. Note that we assume that either the total of investment grants is received at the time of installation of vintage T, or $\gamma$ is defined to represent the present value of these grants over time as a proportion of the price of new capital.
17. We assume that net after tax capital costs are incurred at the end of a period, while the borrowed financial means are made available at the beginning of period T.
18. Using equation (3.3.3.6), we can prove that $\delta_{T,t'} = 1$. The latter implies that the proposed financing of investment in combination with the taxation scheme does not

change the present value of labour costs relative to capital costs: the ratios of the present value of labour costs and the present value of capital costs both before and after taxes are the same. This can be shown using the fact that we have to deal here with a geometric series of the form z , 2zz , 3zzz , . Now, the sum over the first k terms of the geometric series of the form z, zz, zzz, ..., is equal to $S = z(1-z^k)/(1-z)$. We will use this result in deriving the sum S′ of the first k terms of the original series. Note that

$$\begin{aligned} S' = z + z^2 + z^3 + \ldots + z^k + \\ z^2 + z^3 + \ldots + z^k + \\ z^3 + \ldots + z^k + \\ \ldots + z^k + \\ z^k \end{aligned}$$

Therefore, S′ is equal to the sum of k 'normal' geometrical series, the sums of which can be calculated using the expression for S. Indeed, $S' = z(1-z^k)/(1-z) + z^2(1-z^{k-1})/(1-z) + \ldots + z^k(1-z)/(1-z) = z/(1-z) + z^2/(1-z) + \ldots + z^k/(1-z) - k\,z^{k+1}/(1-z) = (1/(1-z))z(1-z^k)/(1-z) - k\,z^{k+1}/(1-z) = z(1-z^k)/(1-z)^2 - k\,z^{k+1}/(1-z)$. Using the expressions for S and S′, it can easily be proved that $\delta_{T,t'} = 1$, since equation (3.3.3.6) consits of the sum of a normal series, S, and the sum of a special series S′.

19. Malcomson, *op. cit.*, p. 28.
20. This implies that the labour coefficient ratio does not change over time. If, however, the influence of disembodied technical progress on labour productivity on new vintages is larger than on old vintages, then the quasi-rent differential is positively influenced by disembodied technical progress, thereby strengthening the condition. We will come back to vintage specific non-embodied technical progress later.
21. Note, however, that the user cost of capital per unit of capacity output increases due to the physical decay of capital once installed. We therefore have to assume that the rate of technical decay is slower than the 'financial rate of decay for tax purposes', in order to arrive at non-increasing user cost of capital per unit of capacity output. The latter assumption may be not that far from reality.
22. Note, however, that there can be circumstances where a rise in real wages leads to scrapping of a fairly large set of vintages due to a possible clustering of old vintages with relatively high labour coefficients that are approximately equal to each other. In this, presumably exceptional, case, a small change in the real wage rate may have dramatic effects on the demand for labour associated with the vintage capital stock. For, the actual value of the effective elasticity of substitution depends primarily on the distribution of the labour coefficients among the vintages which make up the capital stock, as well as the productive capacity of individual vintages. The more skewed this distribution is towards higher labour coefficients, the larger will be the effective elasticity of substitution in case of a rise in real wages, and, in very 'skewed' cases it may even be larger than 1. If the distribution is more or less uniform, which we consider to be the 'normal' case, then the effective elasticity of substitution is presumably smaller than 1.
23. We cannot be sure about this because of the 'putty' character of the ex-ante production function. But it is very improbable that it is not true in actual fact.
24. We will actually calculate the desired rate of labour utilisation using the ex-post information available. We do so by adding capacity labour demand of the vintages which generate the highest quasi-rents per unit of capacity output, such that the associated capacity output adds up to actual output. A production rule for this ex-post optimum set of vintages will be provided in Section 3.6. For a similar application of this principle, see, for instance, Muysken and van Zon, *op. cit.*
25. *Op. cit.*

26. *Op. cit.*, p. 100.
27. *Op. cit.*, p. 101.
28. See Patel, P. and L. Soete (1987), 'The Contribution of Science and Technology to Economic Growth: a Critical Reappraisal of the Evidence', OECD, DSTI/SPR/87.18. They use a perpetual inventory stock construction method with a fixed rate of depreciation.
29. There may, however, be some depreciation in the sense that, as far as a firm or an industry is concerned, R&D stocks can be embodied in people (that may leave) or in licensing agreements (that can be cancelled).
30. If such a kind of complementarity does indeed exist, producers face an interesting problem. They have to allocate resources to physical investment and to 'own' R&D efforts in such a manner that the impact on capital and labour productivity of current and future R&D efforts is 'optimum' from a profit maximisation point of view. Both types of investment decisions should therefore be brought together in an intertemporal framework. In the model on hand, we will allow for the possibility of (partial) complementarity between embodied technological progress and 'own' R&D induced non-embodied technical progress.
31. This is obviously an assumption which underlies the adding up procedure in order to arrive at a measure of the stock of R&D capital. But, especially in the case of R&D (as opposed to physical capital), it seems rather impractical to produce more of the same thing, i.e. applicable and applied knowledge: once a product or process has been developed, there is no use in developing it all over again from the point of view of an individual producer. At a more aggregate level however, R&D expenditures may be allocated to 'imitation' activities, in which case the adding up of R&D expenditures does make sense. But, clearly, the total of R&D expenditures cannot be directed towards 'imitation' activities only, in which case the implicit homogeneity assumption with respect to real R&D expenditures is not fulfilled.
32. *Op. cit.*, p. 101.
33. For an application of the binomial lag structure, see, for instance, Ravenscraft, D. and F.M. Scherer (1982), 'The lag structure of returns to research and development', *Applied Economics*, Vol. 14, pp. 947–67.
34. For n=2 and t=6, for instance, the stock of R&D capital is approximately equal to:

$$M_6 = \sum_{i=3}^{6} U_i \sum_{j=0}^{2} r\&d_{i-j} \cdot W_j$$

This equation can be rewritten as:

$$\begin{aligned} M_6 = \; & U_6.(W_0.r\&d + W_1.r\&d_5 + W_2.r\&d_4) + \\ & U_5.(W_0.r\&d_5 + W_1.r\&d_4 + W_2.r\&d_3) + \\ & U_4.(W_0.r\&d_4 + W_1.r\&d_3 + W_2.r\&d_2) + \\ & U_3.(W_0.r\&d_3 + W_1.r\&d_2 + W_2.r\&d_1) \end{aligned}$$

$$\begin{aligned} = \; & (U_6.W_0).r\&d_6 + (U_6.W_1 + U_5.W_0).r\&d_5 + (U_6.W_2 + U_5.W_1 + U_4.W_0).r\&d_4 + \\ & (U_5.W_2 + U_4.W_1 + U_3.W_0).r\&d_3 + (U_4.W_2 + U_3.W_1).r\&d_2 + (U_3.W_2).r\&d_1 \end{aligned}$$

Or, in more general terms:

$$M_t = \sum_{i=1} r\&d_i \cdot \sum_{j=0}^{\min(i-1,n,t-i)} U_{\min(n,t-i)-j+i} \cdot W_{\min(n,t-i)-j}$$

Because $U_{i-1} = (1 - \delta).U_i$, it follows that the total weight of $r\&d_3$ is smaller than the weight of $r\&d_4$, and the weights of $r\&d_2$ and $r\&d_1$ are smaller than the weight for $r\&d_3$

and furthermore decreasing. Therefore, the lag distribution is skewed towards more recent R&D expenditures. If, however, $\delta = 0$, then the weights $U_i$ for i = 3..6, are equal to 1, and the weight of r&$d_3$ is equal to the weight of r&$d_4$. In addition, the cumulative distribution is symmetric if the binomial distribution is symmetric. It is approximately symmetric, if n is smaller than the half life of the R&D capital stock.

35. Except for the tails of the lag distribution.
36. Note that we assume the constant term in the 'disembodied' technical progress function to be equal to zero for the ex-ante production function. Thus, disembodied technical progress in the ex-ante function refers to quality improvements of products and production processes only. In the ex-post functions we allow for non-zero values of the trends in 'disembodied' technical progress, reflecting 'learning by doing' effects, etc.
37. This switch towards a composite shift function can easily be implemented in the original vintage set-up, as discussed in the previous paragraphs. The only thing we have to do is replace (1 + g) by the product (1 + g).(1 + gn), where gn represents the rate of neutral disembodied technical progress. For an application of such a composite shift function see, for instance, the Freia model of the Dutch economy, as described in Central Planning Bureau, 'FREIA-Een Macroeconomisch model voor de middellange termijn', *Monografie*, 25, Den Haag, 1983; and McHugh, R. and J. Lane (1980), 'The role of embodied technological change in the decline of labour productivity', *Southern Economic Journal*, Vol. 53, pp. 915–26.
38. See, for instance, Central Planning Bureau *op. cit.*, Den Hartog, H. and H.S. Tjan (1976), 'Investment, wages, prices and demand for labour (a clay-clay vintage model for the Netherlands)', *De Economist*, Vol. 114, pp. 32–55, Gelauff, Wennekers and de Jong, *op. cit.*; and Muysken and van Zon, *op. cit.*
39. In this case the R&D sensitivity of the technological progress function will have to be corected for the value of the partial output elasticity of labour.
40. Note that in this context we are only considering 'embodied' technological progress as a measure of the technological distance.
41. That is, constant with respect to the impact on the ex-ante function.
42. If there is (some) R&D complementarity, the influence of embodied technical progress on the age structure of the capital stock will be strengthened by 'own' R&D.

## Acknowledgements

I would like to thank Charles Leedman and George Papaconstantinou of the Directorate of Science, Technology and Industry of the OECD for their comments on an earlier draft of this paper and for trying to improve my English. I would also like to thank the members of the productivity group at MERIT and professor Joan Muysken of the Faculty of Economics at the University of Maastricht for their valuable comments. Last but not least, I am indebted to the Directorate of Science, Technology and Industry of the OECD for their financial support.

# 12. Process innovation and price discrimination

*Patrick J. van Cayseele*

## 1. Introduction

In this chapter, it is shown that the impossibility of separating consumers with a different willingness to pay delays process innovation. It is well known that the *increased* profits resulting from discrimination provide a stronger incentive to engage in *product innovation*, see Bowman (1973). Here, it is shown that *process innovations* are delayed by the innovator's inability to separate consumers. Therefore, the conclusion that the ability to discriminate is favourable to the pursuit of dynamic efficiency is reinforced. This conclusion holds for both durables and non-durables.

For durables, current profits are reduced by the impossibility of separating consumer types. The fact is that consumers willing to buy now anticipate future price reductions, and therefore require 'discounts'. Under these circumstances, buying today is rational since the price charged is sufficiently low to be indifferent between having the good now and having it later, at an even lower price.

Stokey (1979) shows that such 'discounts' are suboptimal. The best policy is to guarantee that prices will never drop in the future. Such a policy can be implemented in the present context by not innovating. If no process innovation takes place, unit manufacturing costs exceed the willingness to pay of low-value customers. Hence, no price reduction can be anticipated. Consumers know that prices will remain at the same level for ever, and hence they will buy today without requiring a discount.

In general, the optimal policy could be to sell to the low-value consumers after a while, that is not to innovate too soon so as to avoid 'big' discounts. The present chapter indicates that such outcomes which are characterised by price discrimination over time are indeed possible. Hence, Salant's (1989) claim is verified that the suboptimality of intertemporal price discrimination in Stokey's model essentially results from the assumption that manufacturing costs remain constant over time. The resulting optimal introduction date falls beyond the one obtained in a setting where inducing self-selection is unnecessary, that is where discrimination can be brought about without discounts.

For non-durables, similar negative 'spillovers' on profitability exist. Once it has become possible to serve the low-value consumers, that is after innovation has taken place, the price will drop to the reservation price of the low-value consumers. Therefore *all* consumers will buy at that price, including the high-value consumers who previously bought at high reservation prices. Again, therefore, *vis-à-vis* a situation in which discrimination can be enforced, innovation will occur at a later date.

In Sections 2 and 3, respectively, the above arguments are presented in a more formal approach. The claims regarding the dynamic inefficiencies that are inflicted upon the economy by the impossibility of separating the consumer groups are proved. In Section 4 a comparison is made in order to determine for which group, the durables or non-durables, one expects this inefficiency to be worst. Section 5 contrasts the present chapter with related work, while Section 6 draws some conclusions.

## 2. Non-durables

An innovator who has a monopoly position (secured perhaps by a patent) faces n consumers who want to buy one unit of his new product at a price v. There are also m consumers who want a unit of the new product at a price w.

The cost of manufacturing the non-durable d is constant over the entire output range. The relation between the above magnitudes is: $v > d > w$; without further process innovation it is simply impossible for the monopolist to serve the low-value consumers.[1]

The monopolist can have such a process innovation at time T at a cost B(T). Without loss of generality, this process innovation reduces the unit manufacturing cost to zero. The innovation cost function B(T) has the following properties: $B' < 0$ and $B'' > 0$, for a discussion see Vislie (1982).

When such a process innovation has taken place, the monopolist can start selling to the low-value consumers. In the absence of discrimination, this implies that the price will drop to w. Profits on the other hand will increase from nv per unit of time up to $(n+m)w$. Clearly it is assumed that:

$$nv < (n + m)\, w \tag{1}$$

in order to make innovation relevant. The net increase in profits per unit of time resulting from innovation therefore is:

$$mw - n\,(v - w) \tag{2}$$

Clearly, as $n(v-w) > 0$, the flow of profits from innovation when the innovator already serves the high-value consumers is lower than the flow of profits when no such high-value consumers exist.

On the optimal timing decision, the replacement of current profits acts so as to delay introduction, as shown in Kamien and Schwartz (1982). More precisely, nv can be seen as the pre-innovation pay-off flow, while $(n+m)\, w$ is the reward for innovation. Kamien and Schwartz show that an increase in $r_0$ *delays* innovation, while an increase in $p_0$ stimulates earlier introduction where $r_0 =$ current profits and $p_0 =$ reward for innovation. In any event, an increase in $r_0$ from 0 (as with no high-value consumers already served) to nv (as with high-value consumers present) will delay process innovation.

## 3. Durables

Again the monopolist faces n high-value consumers who want to pay $V^1$ and m low-value consumers who are ready to pay $V^2$. Again it is assumed that without further process innovation the low-value consumers cannot be served, or $V^1 > c > V^2$. The variable c now denotes the unit manufacturing cost to produce the *durable*. The assumptions regarding the innovation technology are also the same: a process innovation at T reduces unit manufacturing costs to 0 and comes at a cost of B(T), with $B' < 0$ and $B'' > 0$.

When innovation has taken place, the monopolist can serve the low-value consumers. This will make the price charged drop to $V^2$. Therefore, in order to make the high-value consumers buy now, the monopolist must charge a price today which lies below $V^1$. Otherwise the high-value consumers will wait for the process innovation and join with the low-value consumers. Only if the monopolist charges

$$p^1 = (1 - \delta) V^1 + \delta V^2 \tag{3}$$

where $\delta$ is the discount factor associated with a period of length T, that is $\delta = e^{-rT}$ where r is a discount rate, will the high-value consumers self select and buy now. This can be seen from noting that a price given by (3) implies (4):

$$V^1 - p^1 = \delta (V^1 - V^2) \tag{4}$$

which expresses the indifference of a high-value consumer between buying now at $p^1$ (LHS) and buying later at a lower price (RHS).

If the monopolist charges a price given by '(3)', he will sell to all high-value consumers. After the process innovation, the unit costs have fallen to 0 and hence the low value consumers can be served at $p^2 = V^2$.

Therefore, the profit contribution made by innovation is $me^{-rT} V^2$, but in selling to high-value consumers there is a per unit loss given by (4). While it is clear that $me^{-rT} V^2$ decreases in T, (4) shows the opposite as far as profits on the high-value consumers are concerned. Therefore, *vis-à-vis* a situation in which only low-value consumers are served, the optimal introduction date will be delayed as shown in Van Cayseele (1989). This is the dynamic equivalent of the allocative distortions that exist in sorting models, see Cooper (1984).

If one uses an explicit function to represent B(T) explicit introduction dates can also be derived. For example, let

$$B(T) = \frac{r}{2(e^{rT} - 1)} \tag{5}$$

This expression can be derived from an optimal control approach. More precisely, equation (5) is the solution for an innovator who minimises his R&D expenditures while completing an R&D project, see, among others, Van Cayseele (1987). The optimal introduction date then becomes

$$T^* = (-1/r) \ln \{1 - [r/2((m+n)v^2 - nv^1)]^{1/2}\} \tag{6}$$

This clearly shows that in order to make the innovation economic, it has to be the case that $(m+n)V^2 - nV^1 - r/2 > 0$. Otherwise the introduction date will not be finite, a case we will discuss below.[2] If we compare this with the conditions for economic innovation in the non-durable case, noting that a perfectly durable good producing reservation utility v per unit of time yields a reservation value v/r, we see that a similar condition emerges. Indeed, substituting $V^1$ by v/r and $V^2$ by w/r gives the condition $w(m+n) - vn > 0$, up to r/2, a term emerging from the inclusion of the R&D cost function B(T). A precise comparison of introduction dates is given next.

## 4. A comparison between durables and non-durables

Since the impossibility of discriminating delays innovation both in the case of durables and non-durables, it seems appropriate to ask which environment is the most problematic.

This question is tantamount to asking the question in which environment the monopolist loses most from introducing to the low-value consumers. When non-durables are sold, the monopolist loses $n(v-w)$ per unit of time from T on to infinity. The total loss then is

$$L_1 = \frac{n\,(v-w)e^{-rT}}{v} \tag{7}$$

In the case of durables, the monopolist loses on his sales now, however this loss is, as indicated by (4) and multiplied by n

$$L_2 = ne^{-rT}\,(V^1 - V^2) \tag{8}$$

It then should be clear that $L_1 = L_2$ for $V^1$ and $V^2$ are nothing else than respectively v/r and w/r. This can be seen from noting that the reservation value for a high-value consumer who has utility v per unit of time is nothing else than $\int_0^\infty v\,e^{-rT}\,dt = v/r$. Similarly, a low-value consumer who buys at T also extracts utility for an infinite period, but only up to w per unit of time. Hence

$$V^2 = \int_0^\infty e^{-rT}\;wdt = w/r.$$

If it does not make a difference whether durables or non-durables are sold when it comes to the loss that introducing a new product innovation generates, why do we get different introduction dates? In order to answer this question we must focus explicitly on the decision underlying introduction.

For the case of non-durables, introduction is dictated by:

$$\max_T n\,\left(\frac{1-e^{-rT}}{r}\right)(v-d) + (m+n)\left(\frac{e^{-rT}}{r}\right) w - B(T) \tag{9}$$

For durables, on the other hand, introduction dates are obtained from:

$$\max_T n\,[(1-e^{-rT})\,V^1 + e^{-rT}\,V^2 - c] + m\,e^{-rT}\,V^2 - B(T) \tag{10}$$

Clearly, (9) and (10) differ only in the difference in unit manufacturing costs of producing durables versus non durables. Without going into detail, if $(1-e^{-rT})\frac{d}{r} < c$, that is the total manufacturing costs of making non-durables falls short of the one-shot manufacturing cost of producing the service as a durable, then one expects faster introduction of non-durables.

## 5. Related work

Bond and Samuelson (1987) have shown that the producer of a durable could have incentives to innovate even exceeding the socially desirable levels. These results — contrary to ours — are obtained in an environment which does not allow the monopolist to commit himself regarding future sales, or sales dates for that matter. With innovation contractual, that is the innovator cannot change his R&D policy once he has started the project, as well as the necessity to reduce manufacturing costs in order to sell to the low-value consumer, the innovator once he has decided on T is committed. It is simply impossible to sell to the low-value consumers immediately after the high-value consumers have bought.

The relevant comparison, therefore, is with a setting that allows commitment on behalf of the monopolist, see Stokey (1979). There the optimal outcome was never to sell to the low-value consumers. The setting is different from the one here in that the monopolist can serve the low-value consumers at his present costs; that is, there is no innovation needed.

The outcome obtained here is characterised by intertemporal discrimination. The reason is that, unlike in Stokey's model, selling to the low-value consumers can be optimal. The time at which the monopolist starts selling is found by balancing the discounts lost by inducing the high-value consumers to buy now (i.e. make them self-select) and the costs associated with obtaining the necessary process innovation with the benefits of serving the low-value consumers. In Stokey's model, the monopolist faces a continuous demand curve and can charge the one period monopoly price if he decides not to sell to the low-value consumers. If he sells to this group, he gets sales at the low reservation price and has to give a discount to the high-value consumers and, hence, must leave the monopoly price. Clearly, since the monopoly price is optimal, such a policy of increasing sales at low prices cannot be profit increasing. The monopolist will never sell to the low-value consumers in the present model, selling to both consumers can be optimal, as in Varian (1987). The question that has been answered here then is 'when' should one start serving the low-value consumers.

## 6. Conclusion

While it was well known from recent developments in information economics that inducing self-selection leads to allocative inefficiencies or 'distortions', the irrationality of intertemporal price discrimination seemed to indicate that in a dynamic context these problems were absent. The present chapter has proved the contrary. Both for durables and non-durables, process innovation will occur at a slower pace in the absence of opportunities to discriminate. The outcome, however, is one characterised by discrimination, established through price reduction (discounts) and, in view of these discounts, the delay in innovation.

Often in considering innovation the question has been asked whether discrimination should be allowed within the context of a patent, see Hausman and MacKie-Mason (1988). While most of the time the discrimination one has in mind in those models is not of an intertemporal nature, it can be asked whether such intertemporal discrimination can be permitted.

In the present context, intertemporal discrimination will occur in any event. The question then is not whether to allow it (since it implies a price reduction, it is hard to believe there would be opponents), but whether the inability to screen generates costs upon the economy. The answer is yes.

As an extension and topic for further research, one can ask whether correction factors exist which might naturally counteract this dynamic inefficiency. For example, it has been assumed that an initial innovation exists which puts the innovator into a monopoly position. Clearly if rival firms existed in the market (dynamic) competition to serve the low-value consumer i.e. to make the process innovation might eliminate the disincentive, especially since these firms are not serving the high-value consumers. The fact that they innovate sooner only costs money to the monopolist in terms of discounts he has to offer to bribe the high-value consumers into buying now. Finally, in the ex-ante stage, that is where firms compete to innovate first and secure the initial patent, the fact that future process innovations come at the cost of less profits on sales to high-value consumers also will be taken into account.

## Notes

1. Given that process innovation reduces the unit manufacturing costs, some will object to the use of the word 'discrimination' for describing the present situation. Except for trivial cases where the ratio v/d exactly equals (m+n) w/B(T), discrimination in one direction or the other will exist.
2. More precisely, this case can be seen to be Stokey's corner, see Salant (1989). With a finite introduction time, the price profile is a step function:

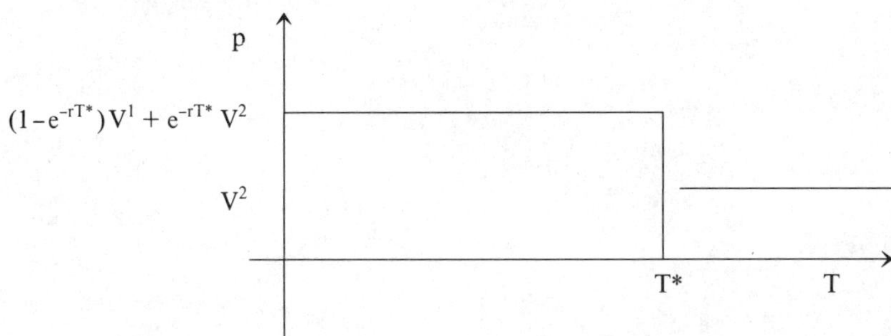


If one introduces many consumer 'types', and in order to serve each group one needs further process innovation, the claim is that for certain distributions of reservation valuations, a sigmoïd price path will emerge. Empirical analyses have documented this behaviour of prices over time.

## Acknowledgements

The author thanks L. Bettendorf, Theon van Dijk, Henri Ergas and Jorge Katz for their comments.

## References

Bond, E. and L. Samuelson (1987), 'Durable goods, market structure and the incentives to innovate', *Economica*, 54, pp. 57–67.

Bowman, W.S. (1973), *Patent and Antitrust Law*, Chicago: University of Chicago Press.

Cooper, R. (1984), 'On allocative distortion in problems with self-selection', *Rand Journal of Economics*, 15, winter, pp. 568–77.

Hausman, J. and J. MacKie-Mason (1988), 'Price discrimination and patent policy', *Rand Journal of Economics*, 19, pp. 253–65.

Kamien, M. and N. Schwartz (1982), *Market Structure and Innovation*, Cambridge University Press.

Salant, S. (1989), 'When is inducing self-selection suboptimal for a monopolist?', *Quarterly Journal of Economics*, 103, pp. 391–7.

Stokey, N. (1979), 'Intertemporal price discrimination', *Quarterly Journal of Economics*, 93, pp. 355–71.

van Cayseele, P. (1987), 'Economies of scope in research and development', *Journal of Economics (Zeitschrift für Nationalökonomie)*, 47 (3), pp. 273–85.

van Cayseele, P. (1989), 'Innovation and Intertemporal Price Discrimination', mimeo, Centrum voor Economische Studiën, K.U. Leuven, p. 5.

Varian, H. (1987), 'Price discrimination', in Schmalensee and Willig, (eds), *Handbook of Industrial Organisation*, Amsterdam: North Holland, (forthcoming).

Vislie, J. (1982), 'A note on an intertemporal cost function for a class of dynamic problems', *Economics Letters*, 8 (3), pp. 215–19.

# 13. Inside the 'green box': on the economics of technological change and the environment

*René Kemp and Luc Soete*

## 1. Introduction

The way in which industrialised countries have produced and consumed goods since the second half of the last century has made unequalled demands on the world's natural environment. Natural resources and sources of energy have been extracted at a great pace; in addition, the environment has had to cope with an enormous amount of waste material. It is now becoming more and more obvious that in the process of rapid technological change and economic growth in the industrialised countries many environmental aspects of new processes and products have been largely disregarded. From an economic point of view, the main reason for this disregard was that environmental damage was not fully assessed and priced. Economic growth and technological development, including a wide range of product, process and organisational innovations, has in many areas moved towards a direction which caused environmental degradation and led to significant costs — either directly or indirectly through a loss of productivity.

The environmental problems of such industrialisation and application of technology, and also the growing awareness at the public and policy level of the environmental costs of present and past growth paths, provide an excellent example of the 'evolutionary' nature of processes of growth and development. The current environmental problems stem from the accumulation of small effects, which at some point in time appear to exceed the critical boundaries of the ecosystem or at least the public perception of those boundaries. They represent a typical example of an *evolutionary* process in which apparently small events, developing in a certain direction during a long period of time, lead to considerable change.

The present environmental problems seem to put limits to past and present paths of industrialisation, transportation and agricultural development. In this chapter we try to provide some insight into the economy–ecology–technology linkages, and especially the technological system, that might help to define and accomplish economic and ecologically sustainable paths.[1] We do this from an evolutionary perspective in which economic growth and technological change is viewed as a complex, non-linear and path-dependent process depending on existing scientific knowledge and technical expertise, supply and demand conditions and infrastructural facilities and institutions. Insight into the evolutionary character of economic growth and technological change is of great importance in designing appropriate policy to achieve sustainable paths of economic and technological development. Policies designed to solve simple linear problems, such as the financing of environmental measures by the government out of the means of economic growth along the same growth paths[2], will be clearly inappropriate.

The structure of this chapter is as follows. In Section 2 we discuss some of the positive and negative 'externality' effects of technological change and the need for an early assessment of these effects. In Section 3 we deal with the interrelation between paths of economic growth and the accompanying networks of 'trajectories' of technological change and the dominance of existing trajectories that might block alternative trajectories that are more environment-friendly. In Section 4 we provide a general economic-theoretical framework for the development and diffusion of pollution abatement technology. And finally, in the last section, we discuss policy issues.

## 2. Technological change and externalities[3]

The problem of positive and negative 'externalities' related to ways of production is relatively well established in the debate surrounding technology policy and in particular technology assessment. The latter has emerged as institutional set up precisely because of the recognition of the 'externality' nature of much technical change. It appeared that a well-informed assessment of new technologies could provide some reassurance about likely impacts or even shape it in a way that is wanted. By definition, though, technology assessment can far better address what we will call here the static distributional aspects of the impact of technology than the dynamic externality aspects of new technologies.

First, from a *static* point of view, and as emphasised in particular by Harvey Brooks, there is an apparent paradox in the impact of technology. The costs or risks of a new technology frequently fall on a limited group of the population, whereas the benefits are widely diffused, often so much that the benefits to any restricted group are barely perceptible even though the aggregate benefit to a large population amounts to considerably more than the total cost to the limited adversely affected group. Examples abound: 'automation', for instance, benefits consumers of a product by lowering its relative price, but the costs in worker displacement are borne by a small number of people, and may be traumatic. A large electricity generating station may adversely affect the local environment, while providing widely diffused benefits to the population served by the electricity produced. Workers in an unusually dangerous occupation such as mining carry a disproportionate share of the costs associated with the resultant materials which may have wide benefits throughout a national economy.

This disproportion between costs and benefits can, however, also work the other way, as in many cases of environmental pollution and emissions. The effluents from a concentrated industrial area such as the Ruhr Valley or the American Great Lakes industrial complex may diffuse acid sulphates over a very large area which derives little benefit from the industrial activity, but which may have its quality of life as well as agricultural productivity seriously degraded thereby.

The issue of sharing costs and benefits of technological change shows how extremely important it is, from both the national and international standpoints: first, to draw up 'rules of the game' to ensure that adverse effects are less harmful than they would be if everything was left to free competition; and, second, to establish such rules fairly early on, before vested interests, acquired privilege and the fierceness of competition jeopardise their compulsory application. Two factors have been noted in recent times which in the applications of new technologies either separately add concern or together might even pose a potential threat: the complexity of systems; and the haste of transfer from an experimental to a market phase.

Contemporary technologies are known to be extremely complex in two respects: they depend increasingly on scientific knowledge and instruments, and further, in order to operate, they assume an organisational fabric which is itself complex. One must indeed speak of systems here and not only of technologies: one is immediately involved in a network of sociotechnical relations involving factors of supply, maintenance, insurance, etc., without which use of the technological product would be impossible. And the more complex the sociotechnical system, the more vulnerable the social organisation is to accident or the obstruction of just one part of the system. At the same time, the knowledge needed to understand the technical operation of the system has become so specialised as to be esoteric to the majority of people. Specialists in increasingly narrow fields have been cut off from each other by their respective skills; to an even greater extent, the multitude of non-technicians has been cut off from the scientists and engineers.

The very scale and complexity of the scientific and technologica enterprise mean that its potential consequences are unprecedented. Whereas the technological risks of the past (pit explosions, railway accidents, dam bursts) can now usually be averted, today's 'major technological risks' threaten larger areas and for a longer time. In the event of a disaster, these areas can no longer be easily isolated and hence evacuated; moreover, toxic emissions (dioxin, radioactive contamination) may have effects that are not detectable for very many years or which last for several generations.

This brings us quite naturally to the increasing importance of *dynamic* 'externality' issues. Here, too, the debate surrounding the introduction of new technologies has addressed many of the issues which at present dominate the debate about the environmental costs and long-term damage. Within a dynamic, evolutionary perspective, long-term externalities are, in Nelson and Winter's words, no longer

> susceptible to definitive once and for all categorization and are more intimately related to particular historical and institutional contexts. To a large extent, the problems involved are aspects of economic change. The processes of change are continually tossing up new 'externalities' that must be dealt with in some manner or other. In a regime in which technical advance is occurring and organisational structure is evolving in response to changing patterns of demand and supply, new nonmarket interactions that are not contained adequately by prevailing laws and policies are almost certain to appear, and old ones may disappear. Long-lasting chemical insecticides were not a problem eighty years ago. Horse manure polluted the cities but automotive emissions

> did not. The canonical 'externality' problem of evolutionary theory is the generation by new technologies of benefits and costs that old institutional structures ignore.[4]

The array of present environmental policy debates about some of the long-term 'externalities' of change, including technological change, in terms of impact on the physical global environment (air, land and water pollution), or even in terms of impact on society's future genetic capital (genetic manipulation, pre-embryo research), or on privacy, are in other words all part of the same need for a continuous 'reassessment' of long-term costs and benefits of change and the accompanying need for institutional adaptation and 'experimentation'. Confronted with an increasing amount of negative environmental externalities of past growth and change, governments are today faced with a major challenge. How to assume the state's function as long-term — as opposed to short-term — social 'regulator' of change in a period not only characterised by an increasingly international environment but also by continual 'new' discoveries — some real, others perceived — of long-term negative environmental externalities of growth and change.

More research on the economy–ecology–technology linkages is clearly needed. Over the last decades, some integrated economic-environmental models have been developed which are evaluated in Nijkamp and Soeteman (1988). They conclude that 'the ecological-economic linkage is indeed still fraught with severe analytical problems. It has to be noted especially that many formal attempts at linking economics and ecology neglect the long-term dynamics of both economic and ecological systems'. Not surprisingly, one of the core research areas they identify in terms of ecologically sustainable economic development is the technological system which has important influences on both the economic and environmental system.

In the next section we deal with the relation between economic growth and clusters of technological trajectories. Within networks of technological trajectories, over time, positive dynamic learning and scale effects can develop which result in substantial welfare gains. On the other hand, the growth of network infrastructure can also lead to serious negative externalities such as congestion. When these negative externalities become more and more important a change of network might be necessary. However, such a change can be costly, especially in the short run, because it takes time for positive externalities to develop.

## 3. Technological networks and growth bifurcations

Starting from a relatively broad economic and technological perspective based on an evolutionary rather than linear line of reasoning — assuming that the future is the result of a relatively unpredictable, complex interaction between economic actors that are endogenous to the process of change, i.e. affected by it, though at the same time affecting and directing it — we now go, albeit briefly, into the interrelation between and direction of economic and technological developments and their relation to the environment. In our view, economic growth, similarly to technological development, is characterised by a non-linear, evolutionary process, typically path-dependent with many bifurcations and possibilities of 'locked-in' development. In terms of some of the terminology now commonly used in the economics of technological change literature, economic growth is likely to be characterised by clusters of economically interrelated technological trajectories, which might give the whole economy growth impulses.

Such clusters of technological trajectories have been identified with new technological systems (Freeman *et al.*, 1982) and new techno-economic paradigms (Freeman and Perez, 1988). The network of technological trajectories related to cheap oil-based energy, combined with mass utilisation of the automobile as a cheap individualised transport system has often been identified with the post-war period of rapid growth. In a similar fashion other networks (e.g. electricity) have been identified with respect to previous periods of rapid growth.

As each system of 'network' infrastructure grows and develops further, more and more negative externalities occur. Congestion, nuisance of all kinds, etc. will gradually increase, so that the growth trajectory will eventually reach its limits. Canals in the eighteenth to nineteenth centuries are a good example, as is horse transport in inner cities at the end of this century. From such a perspective, we would argue that present environmental problems signal in a similar way to earlier congestion problems the limits of the particular growth trajectory. A brief historical analogy might clarify the point we are trying to make.

At the end of the last century the city of London was facing enormous congestion and environmental problems as a result of the use of horses as a means of transport. It is estimated that horses 'produce' no less than 16 k of manure per day. Most street corners in the city of London were manned by so-called crossing sweepers, whose task was not to keep the roads clean, but to clear the way for pedestrians. At the end of the nineteenth century there were about 6,000 crossing sweepers in London. Alternative means of transport had been available for years, but were not used because of restrictive regulations: the red flag amendment, for example, set a speed limit for steam engines of 8 m.p.h. The small-scale production restricted the realisation of dynamic learning and scale effects, and the lack of infrastructural facilities moreover prevented network externalities from arising. In relation to horses, however, cars had a level of about 200 times fewer emissions (measured in grams per mile)*. Whether this caused the eventual disappearance of the horse as means of transport and the rapid development and diffusion of cars (Grübler, 1988) can be left to historians. What we do know is that the growth bifurcation that took place became feasible in environmental terms.

The parallel with today's environmental transport problem is striking. In our view the alternative technological development trajectory is known and has been available for quite some time, i.e. the replacement of car commuter traffic by fully interactive telecommunication systems, allowing for activities like home-work, tele-shopping, home-banking, etc. These alternatives have been available

* Montroll, E.E. and Badger, W.W., 1974. *Introduction to Quantitative Aspects of Social Phenomena*, New York, Gordon and Breach.

in most Western countries for several years as often locally applied technological experiments, but without having much success. Even the ISDN telecommunication system, which has recently been highly praised, has been slow to diffuse and its commercial success not yet established. The reasons for this are similar to those given in the previous example: unforeseen and inappropriate regulations (for example, what is the status of home-workers); the size of the infrastructural facilities required (e.g. initial required costs attached to an ISDN network); the wide range of institutional and infrastructural adjustments and facilities (e.g., the eventual need for a complete transformation of ideas related to the localisation of work; the replanning of cities; the new role of leisure time, etc.).

In other words, in some cases alternative technological 'network' possibilities do exist and are available. They do, however, as in the present case, face major barriers because the positive externalities involved can only develop over time and are prevented from doing so by the existing dominant technological growth trajectory. Their diffusion is in our view essential for an effective growth bifurcation to take place: growth less based on the highly inefficient individualised transport system of persons but on the far more efficient transportation of information.

Besides the mode of transportation, production processes in industry and agriculture also need to be changed. Substantial reductions of emission levels are needed because the present environmental problems stem from accumulation of past emissions, not only from the current flow of emissions. This calls for changes of networks and the economic structure. Pollution control techniques have to be developed, either in polluting firms or in specialised firms, and these techniques have to be implemented throughout the economy. This gives rise to a new economic sector, i.e. the environmental industry, and leads to new and different supplier–user relations. The government plays a crucial role in this process; it has to ensure that the social costs of production are internalised if the market is not capable of doing so. In stimulating the development and spread of innovations in pollution control the government can pursue different policies; it can stimulate technological change in pollution control, either directly by financing or subsidising R&D in pollution control, or indirectly through its environmental policy consisting of instruments of direct control, economic instruments, such as charges and information.

In the literature on environmental economics and the economic literature of technological change, little attention has been paid to the economic and socio-institutional factors that are relevant to the development of innovations in pollution abatement techniques and the factors that influence its diffusion. In the next section we develop an economic-theoretical framework in which this kind of technological change is analysed.

## 4. Factors affecting the development and use of pollution abatement techniques

Both the development of better environmental techniques and changes in the design of products and processes are clearly needed to maintain a sustainable economic development. In the Western world, over the last decades, some pollution abatement techniques, especially with respect to air and water pollution, have been developed and implemented under several acts of government regulation.[5] In this section we deal with the factors that are essential to the development of abatement techniques, both 'end-of-pipe' or 'cleaning' techniques and process-integrated or 'cleaner' techniques, and to the diffusion of such techniques throughout sectors of the economy. We do not deal with households and do not make a distinction between public enterprises and private firms. It will be argued that technological change in pollution control, and especially its diffusion, differs from the traditional process of 'normal' technological change that consists in a succession of new and more efficient production techniques.

Because the factors relevant to the development of innovations in abatement technology are different from the factors that are important for the diffusion of these innovations, they will be discussed separately.

### *Factors affecting the supply of environmental technology*

In dealing with the various factors affecting the supply of technology, we follow Dosi *et al.* (1988) and Cramer and Schot (1989) in making a distinction between (1) technological opportunities; (2) appropriability conditions; and (3) market demand.

*The technological opportunities* with respect to environmental issues differ widely between and within different sectors, from particular available technological opportunities to reduce effluent discharges and emissions to possibilities to decrease the input of certain materials and energy. As in the case of 'normal' technology, these opportunities depend on existing knowledge and equipment, and certain scientific and technical notions. Some environmental problems can be tackled by using available techniques or through factor substitution, while for other environmental problems there is no sufficient technological solution, nor is there expected to be one in the near future.

As in the case of normal technology, the *appropriability conditions* consist of various factors: the time and costs necessary for imitation, statutory protection (patents), the technical advantage over competitors and the extent to which a strong market position can be built up (either through reputation or available distribution channels). Little is actually known about the appropriation conditions of abatement technology. It is clear though that given the public's interest in rapid diffusion of abatement technology, there is probably more government pressure in limiting appropriation than in the case of 'normal' technology. At the same time, increased expectations about regulation and stricter emission controls are likely to lead firms to consider abatement technology as an increasingly important competitive feature.

With respect to *market demand*, firms in the environmental industry have so far faced a very insecure market as a result of the lack of a clear societal demand. The demand for environmental techniques depends on a company's opportunities and willingness to incorporate these techniques in its production process, which, in turn, depends very much on the stringency and nature of the environmental policy. Alternatively, changes in the regulatory framework may undermine the possibilities of firms to develop a sufficient sales potential to engage, for instance, in incremental innovations. On the other hand, the trajectory of progress in terms of abatement goals and aims is pretty stable and well determined.

The above-mentioned factors are all related to the market structure of the environment industry, and, since innovation depends on demand, it also depends on the market structure of polluting sectors.[6] It makes a difference whether abatement technology is developed in regulated firms (through their own R&D activities) or in other firms that specialise in environmental technology. In the former case, innovations in pollution control are often inadequately commercialised as a result of unfamiliarity with the environment market. In the latter case, so-called end-of-pipe or 'cleaning' techniques instead of process-integrated techniques (in which pollution is prevented) will be primarily developed because end-of-pipe techniques offer more opportunities for standardisation in the production process and can be more easily included in the production processes of users (process-integrated techniques often need to be more tailor-made).

As compared to 'normal' technological efforts in industry, pollution control efforts by industry will generally be more focused on incremental improvements in 'cleaning' technology, following — given the uncertainties regarding market demand — relatively well-established technological trajectories of 'progress'. The disparity between best practice and average practice in terms of environmental performance, e.g. emissions, is in general relatively high,[7] thus suggesting the overall dominance of diffusion factors in the analysis of environmental technology.

### *Factors affecting the demand for environmental technologies*

The main reason why new, more efficient production methods are acquired and used, or why new or improved products are brought into production, is because of their supposed contribution to the trading results of a company. In this respect, pollution control innovations differ fundamentally from other types of innovation. Pollution control generally costs money, although this is sometimes compensated for by savings in raw materials and energy (unless taxes are levied on companies' emissions and improvements lead to less pollution costs). Thus, improvements in pollution abatement which are desirable from a social welfare point of view will generally have a negative effect on firms' competitiveness and profits. Although companies might increasingly feel responsible for the damage caused to the environment, cleaner production methods do not represent a priority objective *per se* within companies.[8] As a consequence, companies on their own are unlikely to take action to combat or prevent socially undesirable pollution.[9]

Of the traditional factors affecting the diffusion of particular technologies, the following seem particularly relevant in the case of environmental technology.

*Problems related to knowledge and information.* Apart from being unfamiliar with environmental pollution of industrial activities and products (a problem that is usually under-estimated), a great number of companies, especially small and medium-sized firms, will lack the knowledge to take action against it. This does not only refer to knowledge in the sense of technical expertise that is required to adopt improvements; often, such companies might simply not know what techniques are available, where to turn in order to find out and what forms of technical and financial support they might get. In this respect, the experience companies have had with environmental protection systems is particularly important: environmental problems are recognised sooner and expertise is available on how such problems can be dealt with. In addition, the institutionalisation of environmental protection will have a positive influence on the company's decision process regarding environmental improvements in the production process.

*Insecurity and uncertainty.* Many potential users will be reluctant to adopt cleaner techniques because of the economic risks involved. Production routines and procedures will have to be changed and employees will have to learn and become familiar with the new technology. These economic risks differ between firms due to differences in cost curves. Also, firms have different perceptions about costs and risks. Another form of insecurity is that pollution abatement techniques might become obsolete after some time due to higher standards.

*Supplier–user relationships.* As in the case of the diffusion of 'normal' technology, the diffusion of environmental techniques will crucially depend on the contact between user and supplier. Such relationships are, however, far more difficult to establish in the case of environmental technology. Strong supplier-dependent relations — in Pavitt's sense (1984) — are unlikely to emerge, the user industry being too diverse and covering to some extent the whole industrial and agricultural spectrum. The supplier, on the other hand, will always be delivering a specific technology, i.e. an environmental technique, and therefore, at least so far, not become the major supplier of new technology of the user. The interaction with the using sector, as, for instance, so vividly described in Lundvall's case (1985) of the Danish dairy industry, is therefore less likely to happen.

*Distinction between product and process innovations.* This distinction is particularly important with respect to environmental technology. New products sold in the consumers' market have to meet the needs of consumers, which, in the case of ecological products, depend on whether consumers are environment-minded and prepared to pay for it. The dramatic increase in environmentally 'friendly' goods and design is a good illustration of the perception on the part of industry of the 'sudden' increased environment-mindedness in the public at large. Diffusion of process innovations, on the other hand, as indicated above, depends in the first instance on the company's objectives and values, where cost efficiency occupies a far more prominent position.

As indicated earlier, the diffusion of pollution abatement technology depends on the market structure of the 'polluting' sector. Small firms are less familiar with pollution problems and available techniques to overcome these, and often experience higher costs when adopting abatement technology. Also, the degree of competition between companies and the financial-economic situation is likely to affect the extent to which environmental techniques are adopted. Too much competition and too low profit margins have a clear negative influence on the firm's opportunities and willingness to adopt pollution control methods. Lack of competition might, however, also lead to retarded use and application of environmental techniques if mutual arrangements exist. More research on the relation between market structure and pollution control is needed, however.

To conclude this section, let us recapitulate. Both the development and diffusion of pollution abatement technology is hampered by insecurity and uncertainty about demand, atomistic markets, lack of market structure 'power' of the supplying industry and, above all, exclusion of environmental issues from the firm's traditional objectives and values of profit maximisation. The development and implementation of environmental technology needs, therefore, to be more actively supported than normal production techniques. In designing government policy the factors mentioned in this section should be taken into account. In the next section we address the policy issues.

## 5. Policy issues

It is increasingly understood both by experts and the public at large that the direction of economic growth needs to be changed in order to be ecologically and economically sustainable. Insight into the interrelatedness of paths of economic growth and the accompanying 'trajectories' of technological change and their environmental impact has become of crucial importance in designing appropriate policies. The present environmental problems result from the accumulation of the negative externalities of past economic growth and industrialistion and their impact on 'free' amenities such as air, soil and water. In so far as the environmental problems we are facing now clearly illustrate the limits to growth along the current economic and technological trajectories, they also provide us with hints to possible 'new' directions in which further growth and technological development might be ecologically sustainable.

In the foregoing sections we have outlined some of the factors that affect both the development and the diffusion of environmental techniques, and in what way this differs from 'normal' technological change. The fundamental difference between 'environmental' and 'normal' technological change is that innovations in pollution control (either end-of-pipe techniques or process-integrated techniques) depend on government regulations.

However, the existing technological trajectory of environmental technology is itself also in need of change (see Cramer and Schot, 1989). Up to now mainly cleaning techniques have been developed and implemented. Some of the reasons have already been given in Section 4: end-of-pipe technology can be incorporated in the production process more easily and offers more opportunities for standardisation. The specific 'norm' policy pursued by the government, which is strongly based on available cleaning techniques, has also contributed to the dominance of this 'cleaning' technology trajectory. However, the disadvantage of the application of cleaning techniques is that these techniques in themselves are relatively costly and ineffective (at least in the long run) and often lead to a transfer of environmental problems (e.g. cleaning silt that is polluted by metals). In addition, cleaning techniques are usually aimed at a specific kind of environmental pollution, such as emissions, while other kinds of environmental measures are neglected, for example less, or different, use of raw materials and energy and the treatment or re-use of waste.

It is important that specific environmental aspects be taken into account in the design of processes and products. The change from 'cleaning' environmental technology to process-integrated 'cleaner' technologies, in which environmental damage is prevented instead of taken care of, will not be automatic. Environmental policy should consequently be more oriented towards the development and diffusion of process-integrated techniques, for instance through so-called Pollution-Prevention-Pays Programs as in the US, the improvement of supplier–user relations and especially the use of economic instruments.[10]

The concerns and issues related to the environmental impact of growth and technological advance have suddenly re-emerged in a context very different from that of the mid-1970s, when the issues were first brought to the forefront in the Club of Rome report. To begin with, the evidence on the environmental damage in terms of air, water and soil pollution is by now far more overwhelming. Second, the complexity and time-lags of the interactions between pollution and the ecological system and the surrounding economic and technological environments are still not well, but nevertheless better, understood. Third, the public perception of the environmental problems is far more acute. The hymn to material progress, with the environment to be adapted to the needs and requirements of such progress, appears no longer to be sung with the same conviction. Fourth, at a time when *national* governments are waking up to the importance of the issues involved, it is the *global* dimensions which are most acute. Particularly with respect to environmental issues, the national state, as Daniel Bell suggested, appears indeed too big for the small problems of life and too small for the big problems of life.

## Notes

1. The title of this chapter 'Inside the Green Box' refers to Nathan Rosenberg's seminal book *Inside the Black Box*.
2. A typical example is the announced extra spending on environmental care by the provincial authorities of Noord-Brabant in the Netherlands, due to the more rapid than expected increase in car ownership.
3. We adopt Baumol and Oates' definition of externalities: an externality is present whenever some individuals (say A's) utility or production relationships include real (that is, non-monetary) variables, whose values are chosen by others (persons, corporations, governments) without particular attention to the effects on A's welfare. (Baumol and Oates, 1988, p. 17.)
4. Nelson and Winter (1982), *op. cit.*, p. 368.
5. The government also causes certain environmental problems as in the case of power plants for which pollution control is needed.
6. The influence of the market structure (concentration, competition, etc.) of polluting sectors on demand for pollution abatement techniques will be dealt with later.
7. See e.g. the figures given for two Dutch chemical firms in Reijnders, 1989, p. 904.
8. However environmental awareness in society is growing. Ecology groups, the people living in a company's surroundings and consumers' organisations exert increasing

pressure on a company's environment-mindedness. A bad reputation in this respect may have a negative influence on sales and trading results.

9. Evidently, the willingness to apply cleaner techniques depends on the extent to which this leads to cost increases and lower profits. Sectors facing low profit levels, for example as a result of stagnation tendencies, and sectors operating on the international market, will be less prepared to take measures in this respect.
10. The argument that economic instruments provide a stronger incentive for technological change in pollution control is, however, rather speculative and needs to be investigated more deeply. Here, we do not plan to go into this subject.

## References

Ayers, R.U. and A.V. Kneese (1969), 'Production, consumption and externalities', *American Economic Review*, 59, pp. 282–97.

Baumol, W.J. and W.E. Oates (1988), *The Theory of Environmental Policy*, Cambridge.

Brooks, H. (1982), 'Towards an efficient public technology policy: criteria and evidence', in Giersch, H. (ed.), *Emerging Technologies: Consequences for Economic Growth, Structural Change, and Employment*, Tübingen: J.C.B. Mohr, pp. 329–65.

Cramer, J. and J. Schot (1989), 'Problemen rond innovatie en diffusie van milieutechnologie belicht vanuit een technologie-dynamica perspectief', paper for the RMNO workshop in Utrecht of 12 September, STB-TNO, Apeldoorn.

Dosi, G., Freeman, C., Nelson, R., Silverberg, G. and L. Soete (eds) (1988), *Technical Change and Economic Theory*, London/New York: Pinter.

Downing, P.B. and L.J. White (1986), 'Innovation in pollution control', *Journal of Environmental Economics and Management*, 13, pp. 18–29.

Driel van, P. and J. Krozer (1987), 'Innovatie, preventief milieubeheer en schonere technologie', *Tijdschrift voor Politieke Ekonomie*, 10 (4), pp. 33–56.

Frank, A. and H.J.J. Swarte (1986), *Milieutechnologieën: toepassing in kleine en middelgrote ondernemingen*, Rotterdam: Erasmus Studiecentrum voor Milieukunde.

Freeman, C., Clark, J. and L.L.G. Soete (1982), *Unemployment and Technical Innovation: a Study of Long Waves in Economic Development*, London: Pinter.

Freeman, C. and C. Perez (1988), 'Structural crises of adjustment, business cycles and investment behaviour', in Dosi, G., Freeman, C., Nelson, R., Silverberg, G. and L. Soete (eds), *Technical Change and Economic Theory*, London: Pinter.

Grübler, A. (1988), 'Rise and Fall of Infrastructures, Dynamics of Evolution and Technological Change in Transport', Ph.D. Dissertation, Vienna. Technische Universität.

IJst, P., Stokman, C.T.M. and E.T. Visser (1988), *Informatieoverdracht en informatiebehoefte in de milieuproduktiesector in Nederland*, Zoetermeer: EIM.

Jong, de M.W. and K.J.G. van de Ven (1985), *Milieu-innovaties in kleine ondernemingen; de invloed van het beleid inzake schone technologie*, Economisch Geografisch Instituut/ Stichting Memo, also in Milieubeheer, nr. 21, VROM.

Klink, J., Krozer, J. and A. Nentjes (1989), *Technologische Ontwikkeling door Marktconform Milieubeleid*, conceptrapport voor de NOTA.

Lundvall, B.A. (1985), *Product Innovation and User-Producer Interaction*, Aalborg University Press.

Nationaal Milieubeleidsplan (NMP) (1989), Tweede Kamer, 1988–1989, publ. 21137, nr 1–2, SDU uitgeverij, Den Haag.

Nelson, R.R. and S.G. Winter (1977), 'In search of a useful theory of innovation', *Research Policy*, 6, pp. 36–76.

Nelson, R. and S. Winter (1982), *An Evolutionary Theory of Economic Change*, Cambridge, Mass: Belknap Press of Harvard University.

Nentjes, A. and D. Wiersma (1987), 'Innovation and pollution control', *International Journal of Social Economics*, 15, pp. 51–71.

Nijkamp, P. and F.J. Soeteman (1988), 'Ecological sustainable economic development: key issues for strategic environmental management', *International Journal of Social Economics*, 15, pp. 88–102.

Pavitt, K. (1984), 'Sectoral patterns of technical change: towards a taxonomy and a theory', *Research Policy*, 13, pp. 343–73.

Pearce, D.W. and R.K. Turner (1984), 'The economic evaluation of low and non-waste technologies', *Resources and Conservation*, 11, pp. 27–43.

Quakernaat, J. and J.A. Don (1988), *Naar meer preventie-gerichte milieutechnologie in de industriële produktiesector*, RNMO, nr. 27.

Reijnders, L. (1989), 'De chemische industrie en het milieu', *Economisch Statistische Berichten*, 74 (3724) (13 September 1989).

RIVM, (onder redactie van F. Langeweg) *Zorgen voor Morgen: Nationale Milieuverkenning 1985–2010*, Alphen a/d Rijn, Samson.

Rosenberg, N. (1982), *Inside the Black Box*, Cambridge: Cambridge University Press.

Vollebergh, H. (ed.) (1989), *Milieu en innovatie*, Groningen: Wolters-Noordhof.

World Commission on Environment and Development (WCED) (1987), *Our Common Future*, Oxford: Oxford University Press.

# Index